I0069769

# Understanding Calculus with ClassPad

by

Philip Todd

Saltire Software Inc.,

Beaverton OR

www.saltire.com

Copyright © 2003 by Saltire Software Inc.

ISBN 1-882564-02-2

# Contents

# Introduction

Calculus is typically introduced in a geometrical context. We describe the derivative of a function in terms of the gradient of its tangent. We describe the integral of a function in terms of the area of a collection of trapezoids.

Much of the manipulation involved in the subject, on the other hand, is quite algebraic in nature.

ClassPad is a calculator which is designed to facilitate interchange between graphical, geometrical and algebraic objects. It therefore seemed an interesting project to explore how this tool could be used to illustrate concepts in calculus.

Initially the intention of this book was simply to provide a disjoint set of worked examples using ClassPad, loosely clustered around the common theme of calculus. However, the plan evolved into a more ambitious one: to provide a set of illustrations which in sum cover the skeleton of a basic calculus course.

Why?

Well, clearly nobody is going to replace their favorite calculus text with this one. However, if it is true, as the author believes, that ClassPad is a revolutionary mathematics tool, it may be that some student finds using ClassPad a more congenial way to acquire an understanding of calculus. In which case, it would seem churlish not to give him a fairly complete picture.

This book assumes a basic facility of using the ClassPad calculator. Users are referred to the reference manual or to the book *Getting the Most out of your ClassPad* to attain this facility.

# Differential Calculus

Differential Calculus is about the slopes of tangents to curves.

Why is the slope of a tangent such a big deal? It is easy to get in ClassPad – just create a function, use the tangent tool to draw its tangent and then put its slope in the measurement box – calculus made simple..

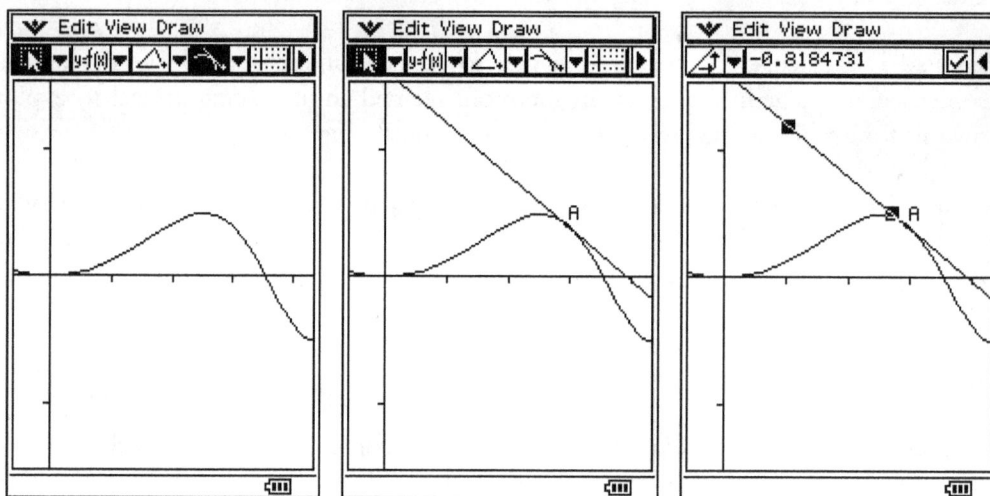

Two answers to the above question:

1. The slope of the tangent is a big deal because it describes the rate of change of the graphed quantity and rate of change is important in the real world. Think about speed of a car (rate of change of position), electrical current (rate of change of charge), work (rate of change of energy) …

2. The slope of the tangent is a big deal because to derive it mathematically, you have to go beyond algebra and work in terms of limits. Specifically the tangent is defined to be the limit of a chord.

In the picture below, the chord AB approximates the tangent at A. As B gets closer to A the approximation gets better and better.

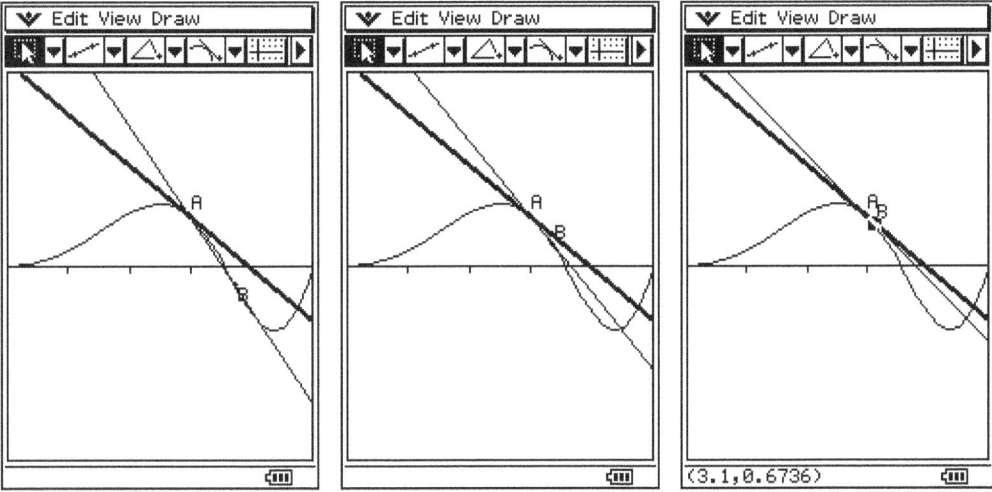

| Edit View Draw | Edit View Draw | Edit View Draw |
|---|---|---|
| A ... B | A ... B | A B |
| | | (3.1,0.6736) |

# Tangents, Slopes and Derivatives

The derivative of a function at a given point is defined as the slope of the tangent to the function at that point. We can illustrate this concept quite handily with ClassPad. Let's take as our function y = sin(2x).

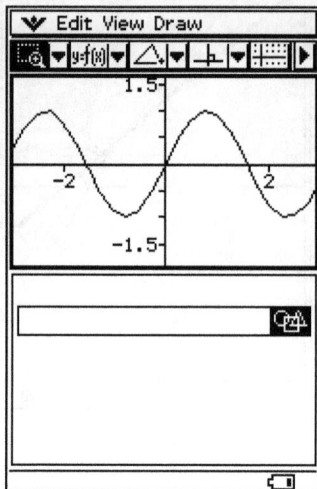

Select the Tangent tool from the construction drop down button and click on the function. This creates a point on the curve and the tangent at that point.

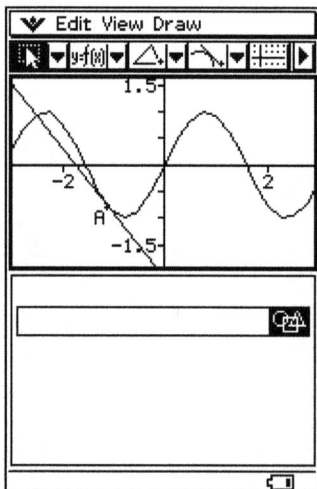

Select point A and the function and create an animation (**Edit/Animate/Add Animation**). We can now watch the tangent traverse the curve by moving the Animation UI slider:

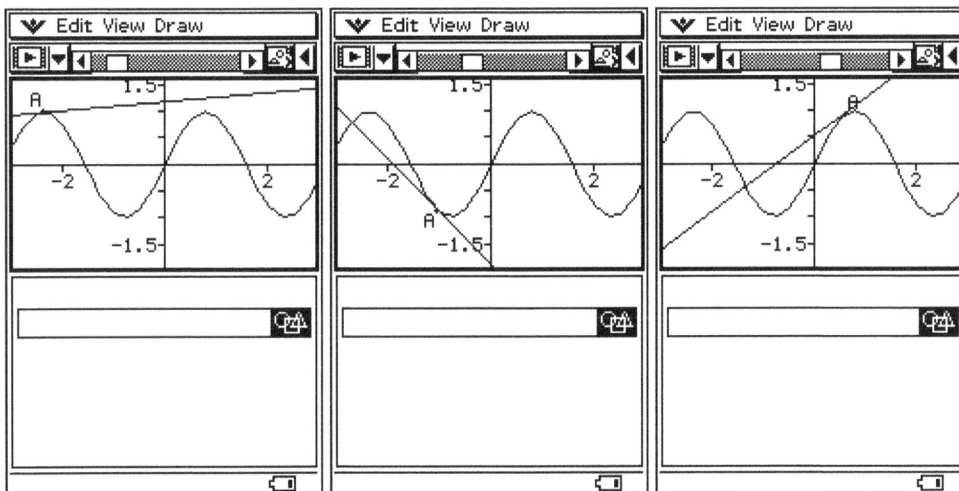

The slope of the tangent can be displayed in the measurement box: select the line and then choose the **Slope** button from the measurement selection drop-down:

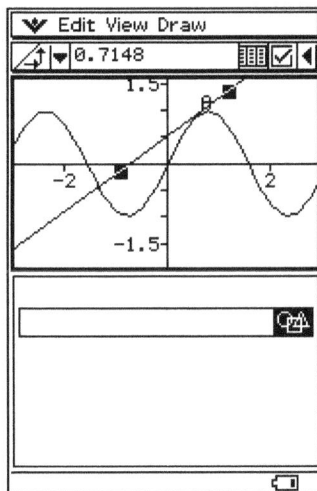

We can tabulate the slope through the course of the motion by clicking the **Table** button to the right of the Measurement box. Notice that the Table button is present if an animation has been defined. Pressing the button tabulates the current measurement over the course of the animation:

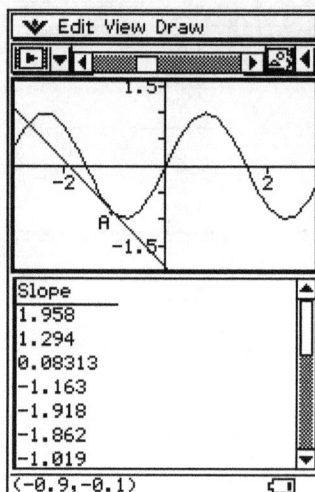

In order to create a graph of the slope over the course of the animation, we'd like to have the x coordinate corresponding to each slope value. Select the point A and display its coordinates in the measurement box, then press the Table button.

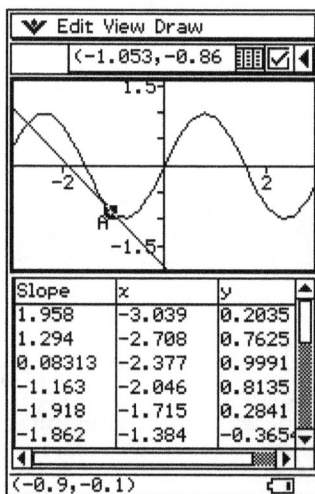

To create a graph of the slope, we need the tables reordered so that x comes before the slope.  Do this by selecting the x column then **Edit / Move to Front** from the menu.

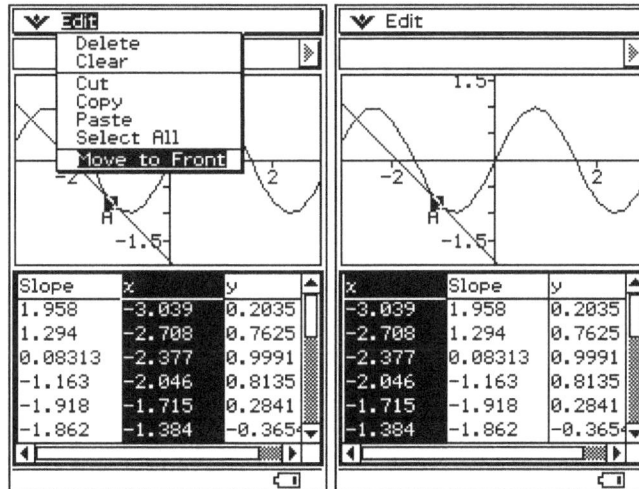

We can now select the two columns x and Slope and drag the pair into the geometry window.  A graph of Slope is created:

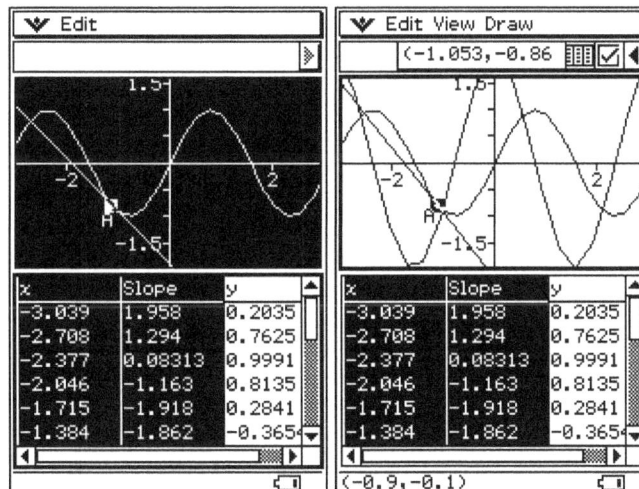

We can illustrate the correspondence between the new function and the tangent by animating a point along the new curve. As the animations run along at the same speed, the point B will illustrate the location on the derived curve corresponding to the slope at A. (You can make the points display larger using **Edit/Properties/Thicker**.)

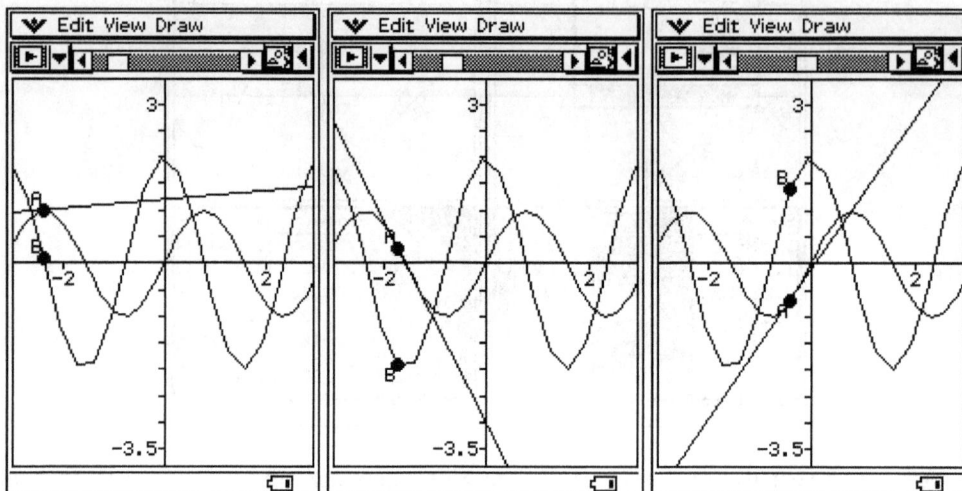

## Name that Function

Can you name the new function? It looks like the original sin(2x), but shifted over so that its maximum is at 0. Let's look at cos(2x):

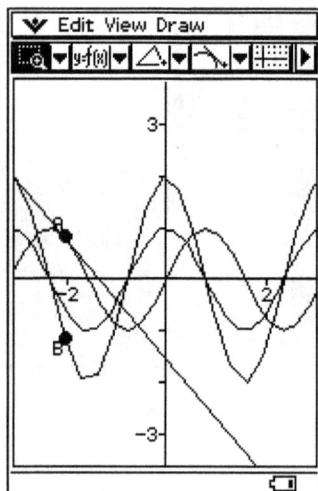

We need to increase the amplitude: let's multiply by 2:

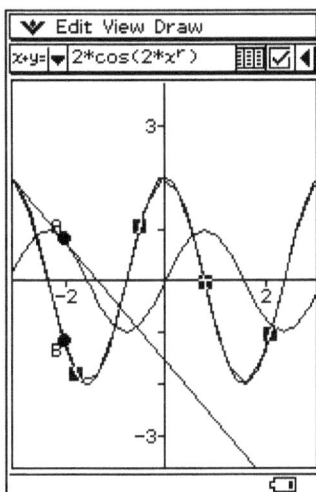

It looks like 2cos(2x) is indeed the function.

EXERCISES

1. Create a plot of the tangent slopes for $f(x) = \dfrac{x^2}{4}$

2. Find a function which describes this plot

3. Create a plot of the tangent slopes for $f(x) = \cos(x)$

4. Find a function which describes this plot

## Chords and Tangents

We'll work with a specific function:

$$y(x) = \frac{x^3}{16} - \frac{x^2}{4} - x + 3$$

Open a Geometry Window inside eActivity and create this function. (Resize and switch on Integer Grid from the **View** Menu. This will aid in accurate location of points.)

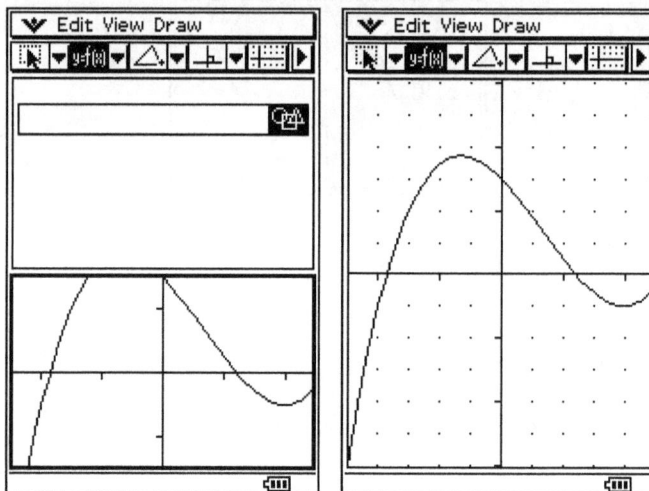

Now create a tangent at x = -2 .

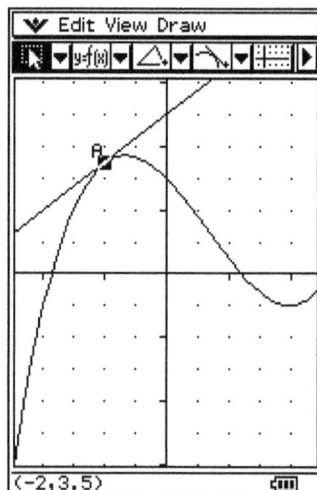

Now you can examine the tangent by selecting it and bringing up the **Measure** Box: You can bring up a display of slope by selecting the slope icon from the Measurement Selection Drop Down.

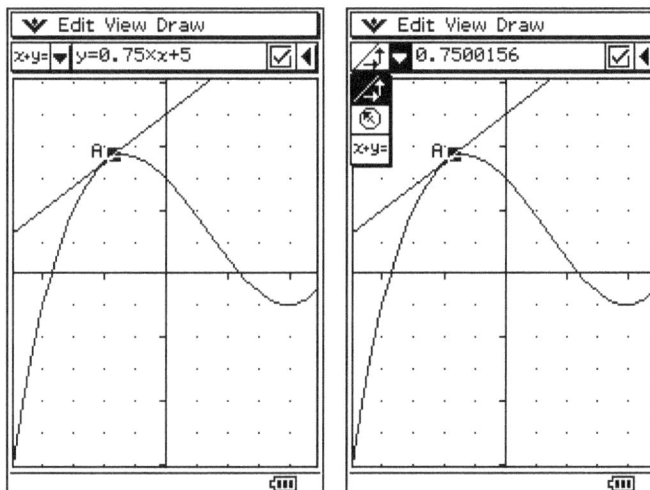

Create a chord through point A using the line tool. Position the second point to lie on the function with it's x-coordinate at –1:

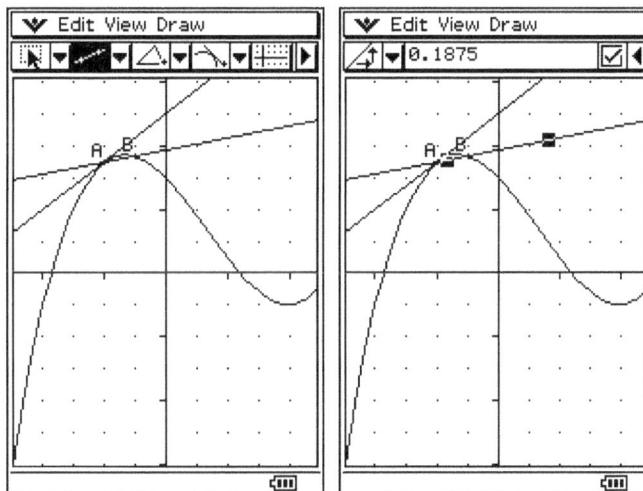

We are going to measure the slope of the chord for different locations of B. In fact, to make life easy for us, we will put B with x-coordinates –1, 0 and 1 (or distance 1,2 and 3 to the right of A). We will collect the chord slope values in a sequence() expression in eActivity. Notice, you can drag and drop the value from the measure box into the expression in eActivity:

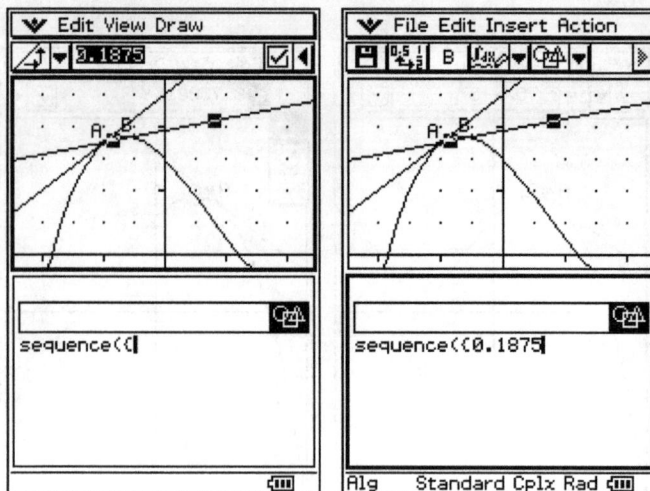

We continue with B located at x value 0:

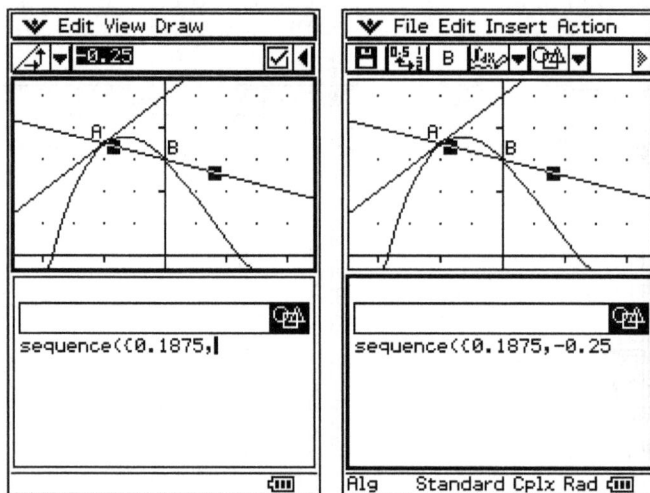

and at x value 1. At this point we can evaluate the sequence (using h as the variable to be consistent with later usage)

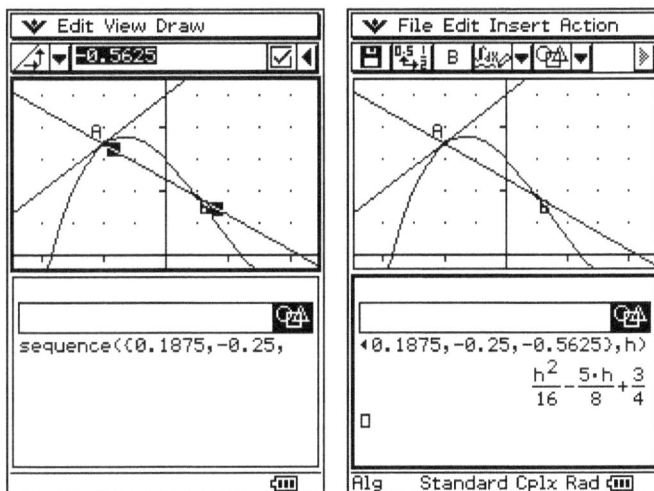

What we get is a quadratic equation which fits these three slope values: when h = 1 it evaluates to the slope of the chord when B is 1 unit to the right of A, when h = 2 it evaluates to the slope of AB when B is 2 units to the right of A. When h = 3 it evaluates to the slope of AB when B is 3 units to the right of A.

What do we get when h = 4? Drag and drop the quadratic into a new input line and find the vertical bar on the MATH/OPTN Keyboard:

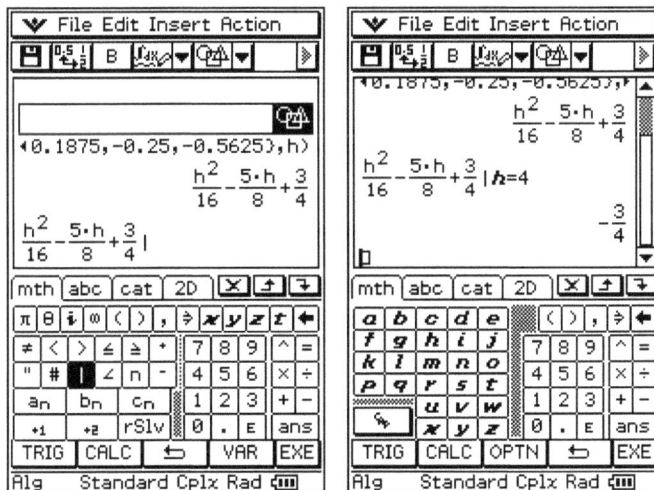

Now drag B so that its x coordinate is 2 and examine the slope of the chord, we see that it corresponds to the formula:

```
 ▼ Edit View Draw
 ◢ ▼ -0.75                    ☑ ◀
◀0.1875,-0.25,-0.3625,▸ ▲
                      h²   5·h   3
                      ── - ─── + ─
                      16    8    4
 h²   5·h   3
 ── - ─── + ─ | h=4
 16    8    4
                             3
                           - ─
                             4
 □                            ▼
```

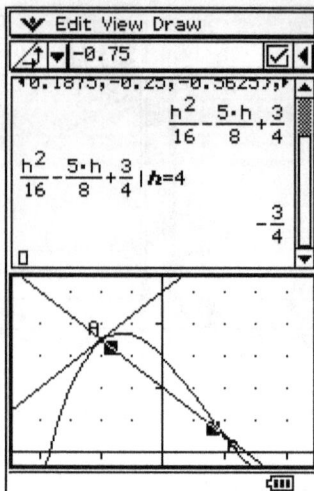

Now what becomes of our formula, when h is 0?

```
 ▼ File Edit Insert Action
 🖫 ⁰·⁵₊²  B  📐▼  ◷▲▼        ▷
◀0.1875,-0.25,-0.3625,▸ ▲
                      h²   5·h   3
                      ── - ─── + ─
                      16    8    4
 h²   5·h   3
 ── - ─── + ─ | h=0
 16    8    4
                             3
                             ─
                             4
 □                            ▼
Alg    Standard Cplx Rad ▥
```

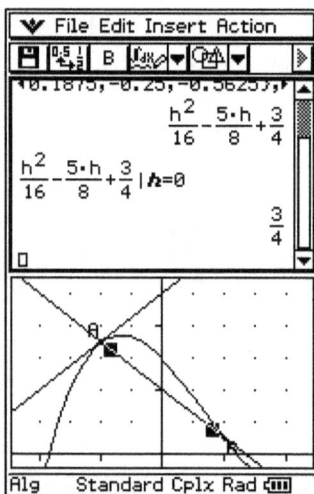

It evaluates to its constant term ¾, which is the gradient of the tangent at A.

CAVEAT

Please note that this example was pre-rigged so that the sequence command with 3 points would give an exact formula for the equation of the chord. This will only work if the input function is a cubic. You'll see why in the next section, on chord algebra.

EXERCISES

1. Find a formula for the slope of the chord to the function $y = x^2$ between the points x=0 and x=h.

2. Find a formula for the slope of the chord to the function $y = x^2$ between the points x=1 and x=1+h.

3. Find a formula for the slope of the chord to the function $y = x^2$ between the points x=2 and x=2+h.

4. Can you find a pattern. Verify this algebraically.

## The Algebra of Chord Slopes

We remember that the slope of a line between two points is the difference in the y coordinates divided by the difference in the x coordinates. Let's assume point A and B both lie on the curve y=f(x), and let's assume A is the point (x , f(x)) and B is the point (x+h , f(x+h)). Then the slope of the chord is:

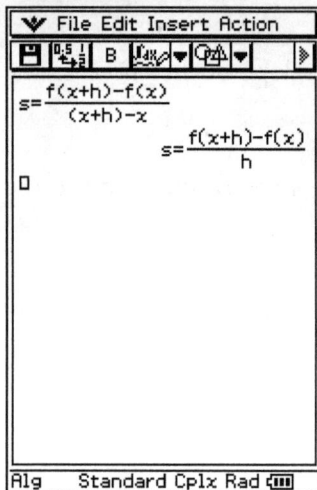

```
▼ File Edit Insert Action
[icons]  B  [icons]
```

$$s = \frac{f(x+h) - f(x)}{(x+h) - x}$$

$$s = \frac{f(x+h) - f(x)}{h}$$

□

```
Alg    Standard Cplx Rad
```

Now let's use the same curve as before:

$$f(x) = \frac{x^3}{16} - \frac{x^2}{4} - x + 3$$

We can enter the curve equation into ClassPad, then use a little copying and pasting and editing to create the chord equation (remembering parentheses where necessary):

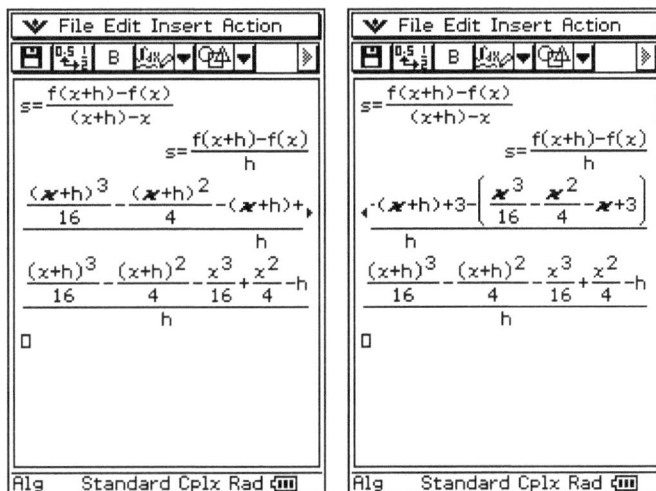

This is a bit messy, but we can apply expand():

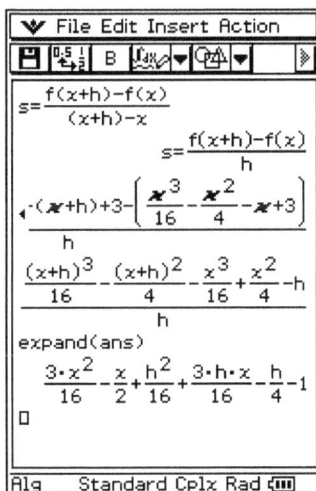

And if we evaluate the expression at x = -2, we can get an algebraic equation for the chord slope (in fact the same equation as we got using the sequence() command in the above section):

Alternatively, if we evaluate the expression for the chord slope where h = 0, we get an equation for the tangent at point x. Evaluating this at x=-2 gives the slope of the tangent at A:

## Limits

The limit function gives us a way to evaluate an expression which we might otherwise shrink from evaluating , for example:

$$f(x) = \frac{x}{(x-2)^2} \text{ at x} = 2$$

$$f(x) = \frac{\sin(x)}{x} \text{ at x} = 0$$

Let's examine the second expression. An attempt to evaluate it at x=0 fails because both the numerator and denominator of the expression are 0:

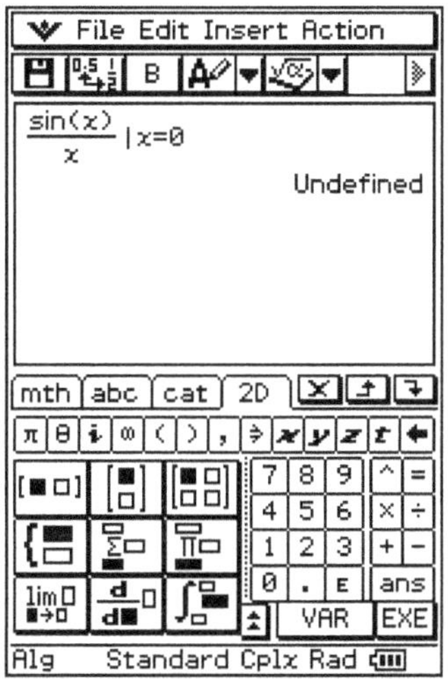
File Edit Insert Action
sin(x)/x |x=0
Undefined
mth abc cat 2D
Alg Standard Cplx Rad

How about evaluating at a small value for x (converting to decimal using the **exact/approx** button).

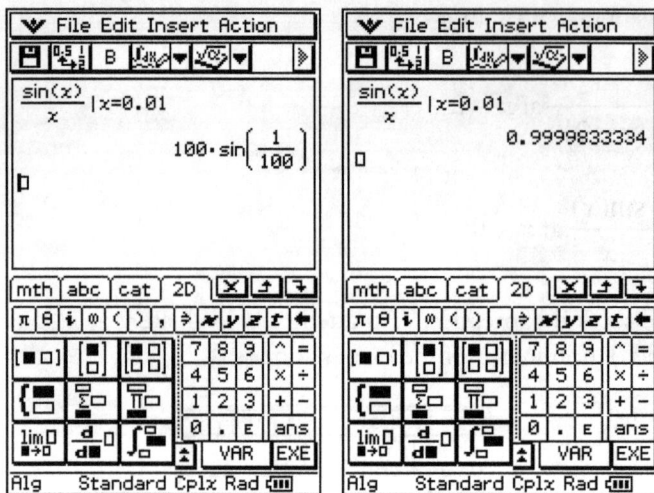

As we make x smaller and smaller the value appears to get closer and closer to 1:

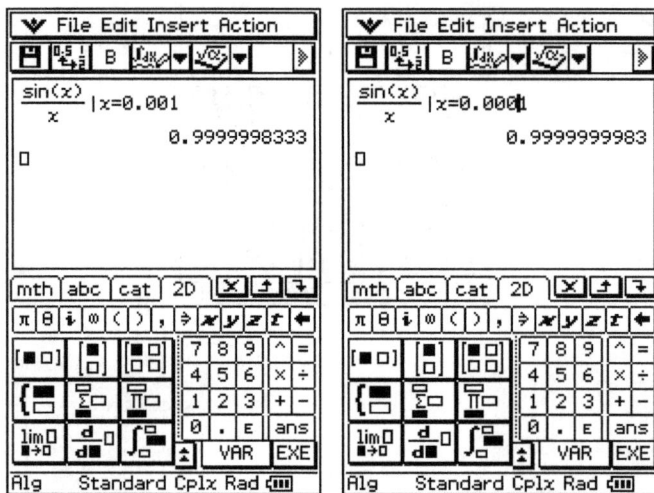

To find the actual limit, we invoke the limit function from the second pane of the 2D Math keypad:

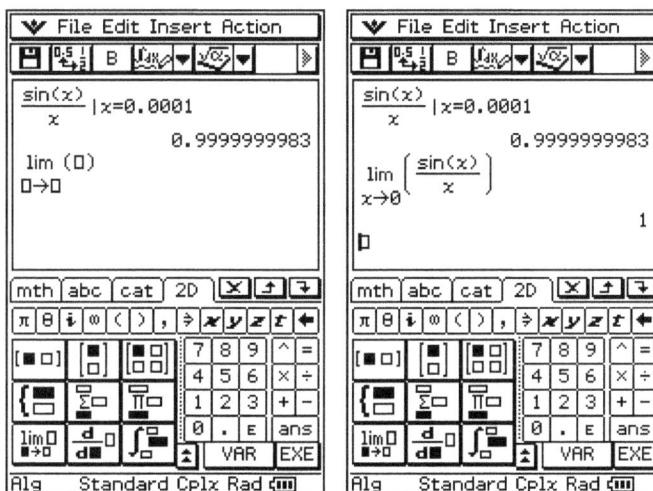

Other limits we'll use later are:

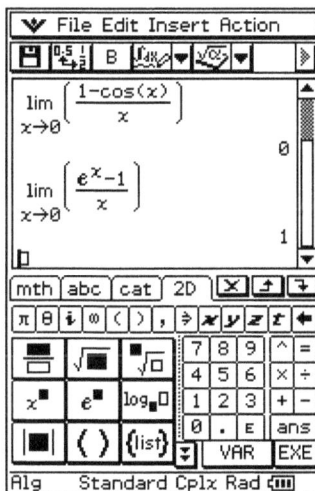

## Limits and Tangent Slopes

We saw before that the slope of the chord through the points (x,f(x)) and (x+h,f(x+h)) is:

$$\frac{f(x+h) - f(x)}{h}$$

The slope of the tangent at x is the limit of this expression as h tends to 0:

$$\lim_{h \to 0} \frac{f(x+h) - f(x)}{h}$$

ClassPad has the limit function. Let's use it to examine the slopes of the tangents for some simple functions:

| | | |
|---|---|---|
| ▼ File Edit Insert Action | ▼ File Edit Insert Action | ▼ File Edit Insert Action |
| $\lim_{h \to 0}\left(\frac{(x+h)^2 - x^2}{h}\right)$ | $\lim_{h \to 0}\left(\frac{(x+h)^3 - x^3}{h}\right)$ | $\lim_{h \to 0}\left(\frac{(x+h)^4 - x^4}{h}\right)$ |
| $2 \cdot x$ | $3 \cdot x^2$ | $4 \cdot x^3$ |
| mth abc cat 2D | mth abc cat 2D | mth abc cat 2D |
| Alg  Standard Cplx Rad | Alg  Standard Cplx Rad | Alg  Standard Cplx Rad |

Do you observe a pattern?

Confirm it by entering in the general form:

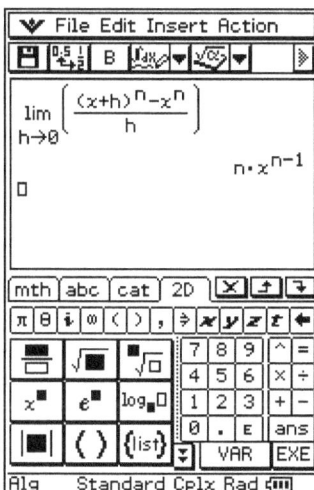

$$\lim_{h \to 0} \left[ \frac{(x+h)^n - x^n}{h} \right]$$

$$n \cdot x^{n-1}$$

## Tangent Slope from First Principles y=xⁿ (n positive integer)

In the above examples, ClassPad's limit function is doing the hard work for us. Let's see what progress we can make with these expressions without resorting to the heavy equipment.

For a start, let's look at the function $f(x) = x^2$. First let's expand out f(x+h), then construct the formula $\dfrac{f(x+h) - f(x)}{h}$

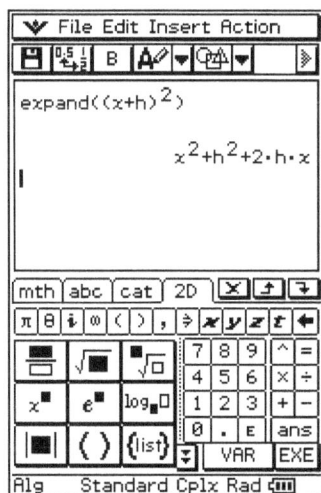

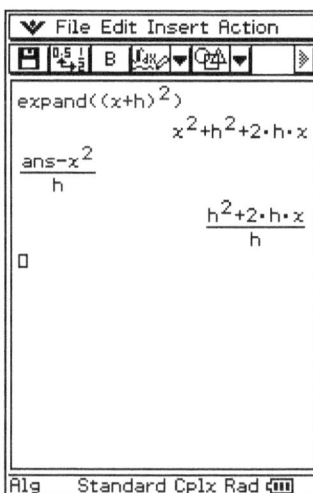

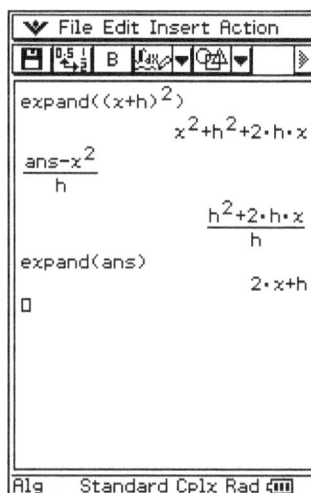

expand((x+h)^2)
$$x^2 + h^2 + 2 \cdot h \cdot x$$

expand((x+h)^2)
$$x^2 + h^2 + 2 \cdot h \cdot x$$
$$\frac{ans - x^2}{h}$$
$$\frac{h^2 + 2 \cdot h \cdot x}{h}$$

expand((x+h)^2)
$$x^2 + h^2 + 2 \cdot h \cdot x$$
$$\frac{ans - x^2}{h}$$
$$\frac{h^2 + 2 \cdot h \cdot x}{h}$$
expand(ans)
$$2 \cdot x + h$$

The expansion of $(x+h)^2$ has a term in $x^2$, a term in h and a term in $h^2$. We notice that the $x^2$ disappears when we form the numerator, and that the h disappears from the term 2.x.h when you divide by the denominator.

Evaluating at h = 0 gives the expression for the tangent slope.

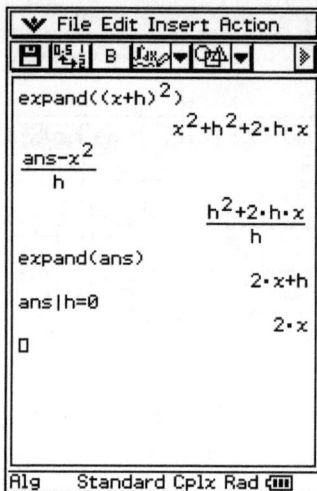

```
 ▼ File Edit Insert Action
 💾 ⬚ B ⬚▼⬚▼        ≫

expand((x+h)²)
                   x²+h²+2·h·x

 ans-x²
 ─────
   h
                    h²+2·h·x
                    ────────
                       h
expand(ans)
                      2·x+h

ans|h=0
                        2·x

 □


 Alg    Standard Cplx Rad ▥
```

We can see that the critical term is the term in h in the original expansion.

We can repeat the analysis for the function $f(x) = x^3$ simply by changing the 2 to 3 in the second expression, then in the first expression, then press EXE (while the cursor is in the first equation) this will cause the whole sheet to re-evaluate.

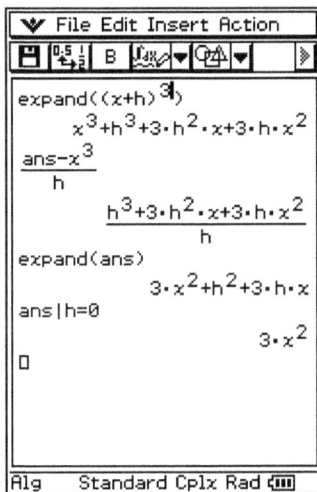

```
▼ File Edit Insert Action
🖫 📲 B  ▦▾ 🖾▾        ⟫
expand((x+h)³)
        x³+h³+3·h²·x+3·h·x²
ans-x³
 h
          h³+3·h²·x+3·h·x²
          ─────────────────
                 h
expand(ans)
          3·x²+h²+3·h·x
ans|h=0
                    3·x²
□

Alg    Standard Cplx Rad ▥
```

We observe that for these two cases, the critical term in h in the expansion of $(x+h)^n$ is:

$$nx^{n-1}h$$

Is it generally true? We can prove it by induction on n. The above shows that it is true for n=2 and 3, let's assume it is true for some general n, and show that the result holds for n+1:

First we write the expansion of $(x+h)^n$ where we cluster all the terms in powers of h higher than 1 into a single term $uh^2$. (u itself is a polynomial in x and h.) Then we multiply this by (x+h). Some manipulation extracts the h term in the new expression:

We can drag the coefficient of h into a new expression and massage it into a nicer form:

Which proves the result.

## Tangent Slope from First Principles y=xⁿ (n negative integer)

So much for positive integer powers, how about negative integer powers?

First let's take n = -1:

```
▼ File Edit Insert Action
┌─┬─┬─┬───┬───┬─────┐
│💾│⁰·⁵│B│   │   │     │  ≫
└─┴─┴─┴───┴───┴─────┘
combine( 1/(x+h) - 1/x )
                        -h
                     ────────
                     x·(x+h)

ans
───
 h
                        -1
                     ────────
                     x·(x+h)

ans|h=0
                        -1
                       ────
                        x²

▯

Alg    Standard Cplx Rad ▭
```

We see that applying the combine() function to the numerator of our chord equation, then dividing through by h yields the result $\dfrac{-1}{x^2}$, which can be rewritten $-1 \cdot x^{-2}$, which agrees with the general form $n \cdot x^{n-1}$ derived above.

We can try the same analysis with $x^{-2}$ (This time we need to apply the simplify() function to persuade the h to cancel).

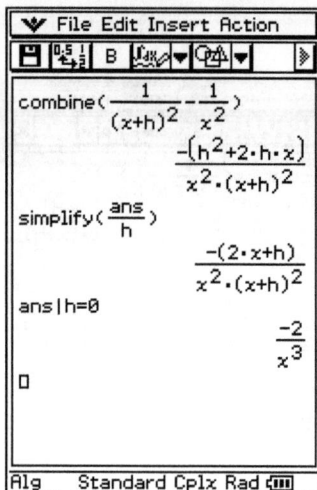

```
▼ File Edit Insert Action
────────────────────────────────
[□][0.5][ B ][≈▽][⚙▽]        [▷]
────────────────────────────────
combine( ─────── - ── )
         (x+h)²    x²
                       -(h²+2·h·x)
                       ───────────
                        x²·(x+h)²
simplify( ans )
          ───
           h
                        -(2·x+h)
                        ────────
                        x²·(x+h)²
ans|h=0
                           -2
                           ──
                           x³
□

────────────────────────────────
Alg    Standard Cplx Rad ▭
```

Again this agrees with the general form $nx^{n-1}$.

Dare we attempt this analysis for general negative integer power n?  With ClassPad's help, of course we do  First we combine the expression on the numerator of the chord slope expression:

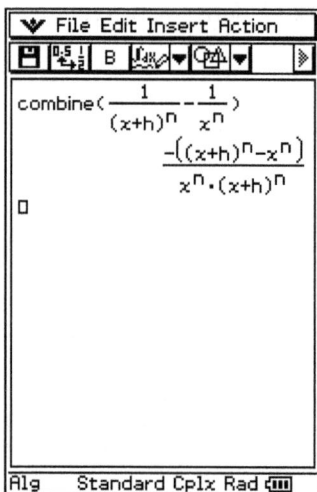

```
▼ File Edit Insert Action
────────────────────────────────
[□][0.5][ B ][≈▽][⚙▽]        [▷]
────────────────────────────────
combine( ─────── - ── )
         (x+h)ⁿ    xⁿ
                     -((x+h)ⁿ-xⁿ)
                     ─────────────
                      xⁿ·(x+h)ⁿ
□






────────────────────────────────
Alg    Standard Cplx Rad ▭
```

Now we need to expand this, but ClassPad cannot expand $(x+h)^n$

We can, however, in the same way as above (lumping all the terms in powers of h higher than 1 into a single term $uh^2$).

```
File Edit Insert Action
combine( 1/(x+h)^n - 1/x^n )
                 -((x+h)^n - x^n)
                 ─────────────────
                   x^n·(x+h)^n
 -((x^n + n×x^{n-1}xh + uxh^2) - x^n)
 ────────────────────────────────────
            x^n·(x+h)^n
            -(h^2·u + h·n·x^{n-1})
            ──────────────────────
                 x^n·(x+h)^n

Alg    Standard Cplx Rad
```

Now divide through by h, simplify, and set h to 0:

```
File Edit Insert Action
combine( 1/(x+h)^n - 1/x^n )
                 -((x+h)^n - x^n)
                 ─────────────────
                   x^n·(x+h)^n
 -((x^n + n×x^{n-1}xh + uxh^2) - x^n)
 ────────────────────────────────────
            x^n·(x+h)^n
            -(h^2·u + h·n·x^{n-1})
            ──────────────────────
                 x^n·(x+h)^n
simplify( ans/h )
                 -(h·u·x^{-n+1} + n)
                 ───────────────────
                     x·(x+h)^n
```

```
File Edit Insert Action
 (x+h)^n   x^n
                 -((x+h)^n - x^n)
                 ─────────────────
                   x^n·(x+h)^n
 -((x^n + n×x^{n-1}xh + uxh^2) - x)
 ──────────────────────────────────
            x^n·(x+h)^n
            -(h^2·u + h·n·x^{n-1})
            ──────────────────────
                 x^n·(x+h)^n
simplify( ans/h )
                 -(h·u·x^{-n+1} + n)
                 ───────────────────
                     x·(x+h)^n
ans|h=0
                       -n·x^{-n-1}

Alg    Standard Cplx Rad
```

EXERCISES

1. What is the slope of the tangent to the curve $f(x) = x^2$ at x=2. Verify using ClassPad's Geometry application.

2. What is the slope of the tangent to the curve $f(x) = \dfrac{1}{x}$ at x=-1. Verify using ClassPad's Geometry application.

3. Write down an expression for the slope of a chord of the function $f(x) = \dfrac{1}{x^2}$. From this, derive an expression for the tangent slope.

## Derivatives by First Principles

Another name for the slope of the tangent of a curve f(x) at x is the ***derivative of f with respect to x.*** This can be written:

f'(x)

Or alternatively:

$$\frac{df}{dx}$$

ClassPad uses the latter notation:

Another name for finding the derivative of f is ***differentiating*** f.

## Derivative of sin() and cos()

Let's go ahead and try differentiating some familiar functions:

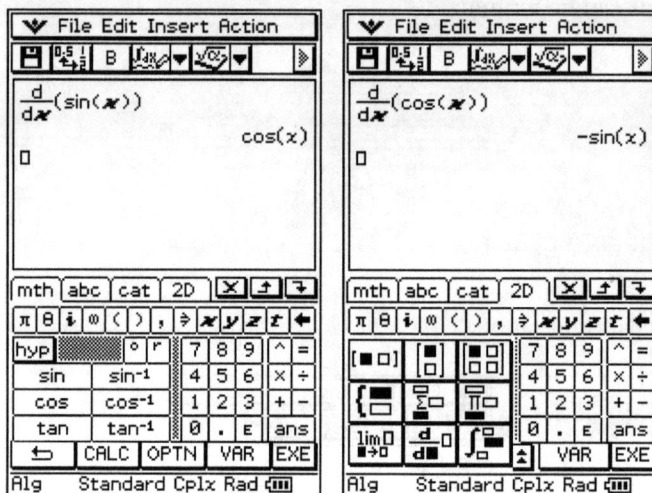

Can we use ClassPad to derive these from "first principles" using the expression for the slope of a chord:

Let's try sin(x). Our first inclination is to apply some trig manipulation, tExpand() should do the trick.

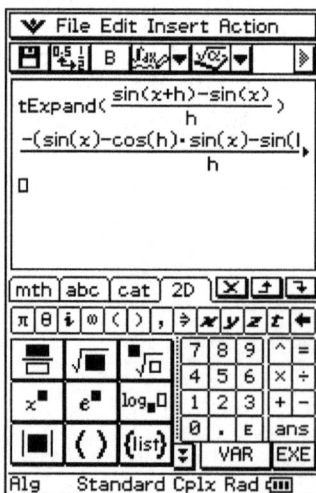

Now unfortunately the sin(x) term has not disappeared. The best we can do is collect the two sin(x) terms together:

```
W  File Edit Insert Action
[icons]
tExpand( sin(x+h)-sin(x) )
            h
 -(sin(x)-cos(h)·sin(x)-sin(!
              h
collect(Ans,sin(x))
( 1 ·cos(h)- 1 )·sin(x)+ 1 ·sir▸
  h         h           h
□



Alg   Standard Cplx Rad [iii]
```

Let's examine the coefficient of sin(x):

```
W  File Edit Insert Action
[icons]
                     -sin(x)
tExpand( sin(𝒙+h)-sin(𝒙 ,
              h
 -(sin(x)-cos(h)·sin(x)-sir ,
              h
collect(ans,sin(𝒙))
( 1 ·cos(h)- 1 )·sin(x)+ 1 ·:▸
  h         h           h
combine( 1 ·cos(h)- 1 )
         h          h
                    cos(h)-1
                       h
lim ( cos(h)-1 )
h→0      h
                         0
□
Alg   Standard Cplx Rad [iii]
```

This combines into the expression $\dfrac{\cos(h)-1}{h}$, whose limit we have seen before is 0.

We now need to examine the coefficient of cos(x):

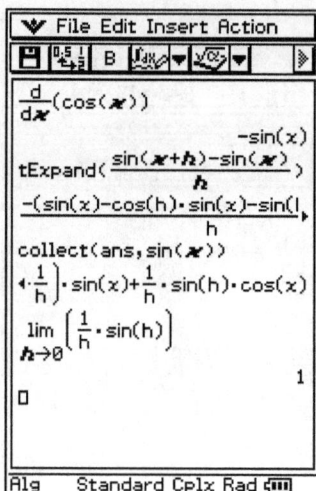

```
▼ File Edit Insert Action
🖫 📊 B 📊▼ 📇▼          »

d
──(cos(𝒙))
d𝒙
                        -sin(x)
       sin(𝒙+𝒉)-sin(𝒙)
tExpand(──────────────,
              𝒉
-(sin(x)-cos(h)·sin(x)-sin(⌐
            h
collect(ans,sin(𝒙))
    1                1
◄·───)·sin(x)+───·sin(h)·cos(x)
    h                h
      ⎛ 1         ⎞
lim   ⎜───·sin(h) ⎟
𝒉→0   ⎝ h         ⎠
                              1
□

Alg    Standard Cplx Rad ▥
```

Again this is an expression we have seen before: $\dfrac{\sin(h)}{h}$, and its limit is 1.

To recap, the coefficient of sin(x) tends to 0 whereas the coefficient of cos(x) tends to 1. Hence the limit of the combined expression is cos(x).

# Exponentials

In the ClassPad Geometry create the function $e^x$. You'll need to zoom and pan to get it to fit nicely onto the screen:

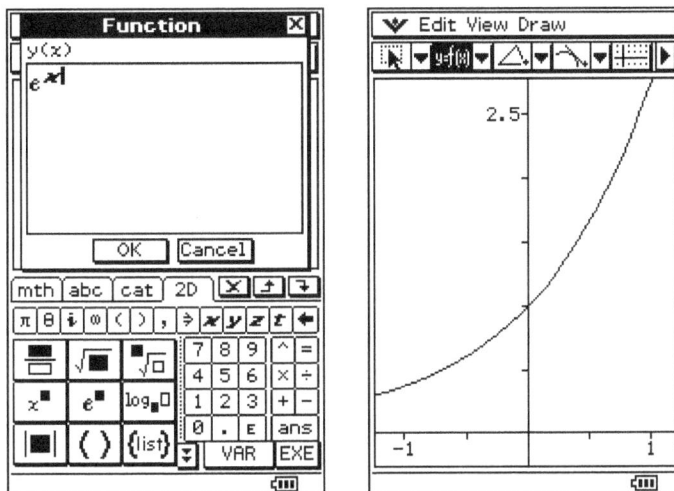

Now create a tangent to the curve:

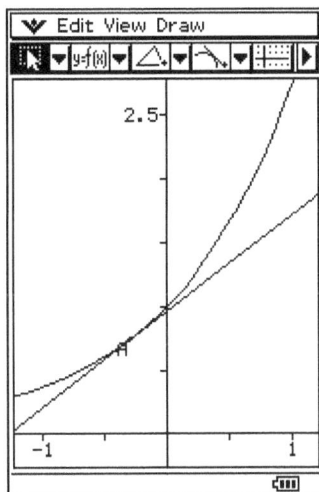

The slope of this tangent is the derivative of the curve evaluated at the point A.

Examine the slope of the tangent, and compare it with the y-coordinate of point A. What do you notice?

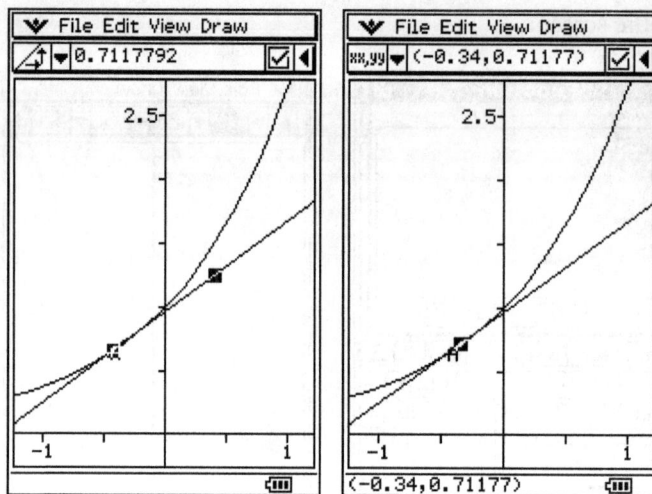

To look at the derivative of $e^x$, however, we'd like to see the slopes of the tangents at many points on the curve. We can do this in ClassPad by animating the point A and collecting the slopes as A slides along the curve.

First select A and the curve and the menu item **Edit/Animate/Add Animation**.

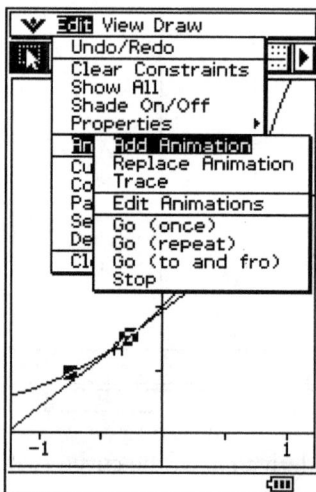

This creates an animation. You can watch the tangent change as A slides along the curve:

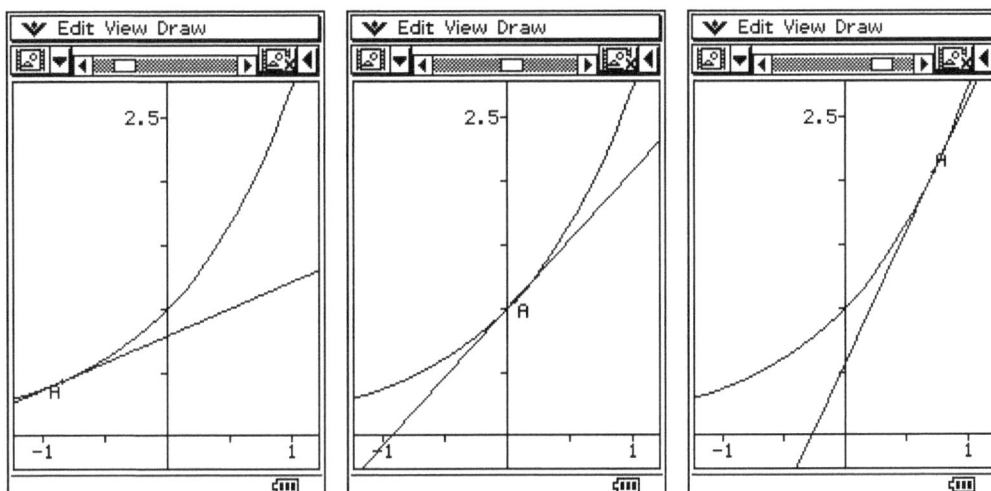

Now, we can examine the slope of the tangent by selecting it and bringing up slope in the Measure Box:

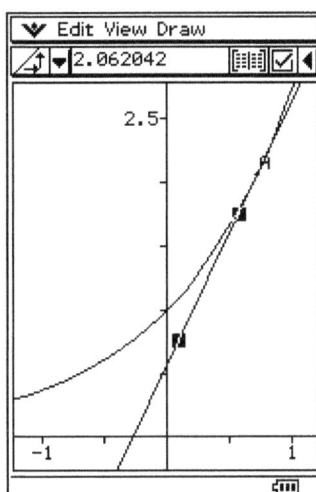

To create a table of slope values, we can tap on the **Table** button

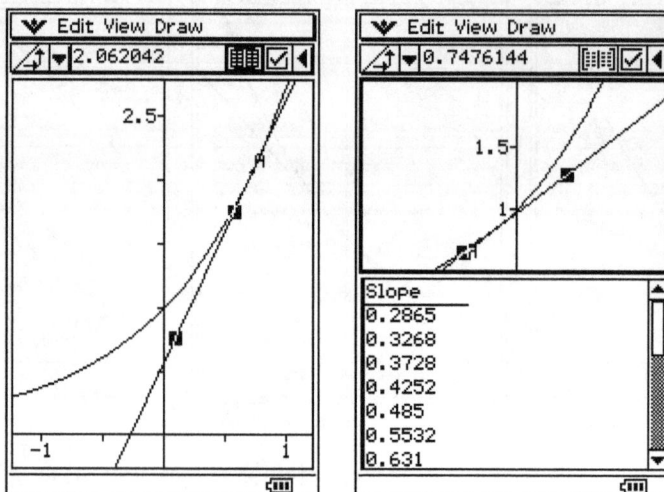

Now, we'd like to plot slope against x coordinate of A. To do this we need to tabulate the x coordinates. Select A, make sure the coordinates are displayed in the Measure Box, then click the Table button:

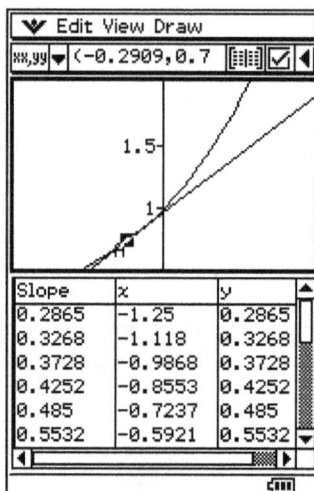

To display a plot of coordinate pairs (x,Slope), all we need to do is drag these two columns up to the graph – however, first we have to make sure the x column in in front of the slope column (otherwise we'd get a graph of (Slope,x))

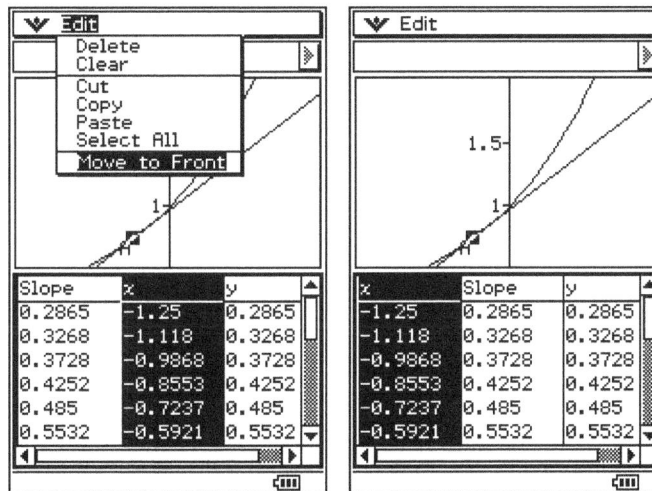

Now, we select the two columns and drag and drop:

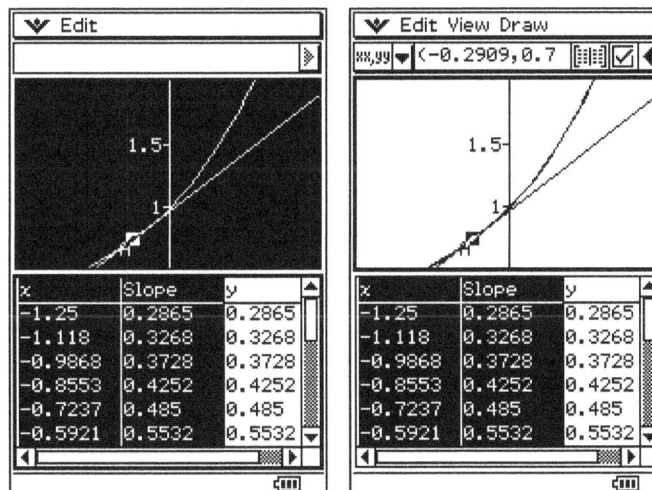

You can see from a faint thickening of the lines that the curve has dropped right on top of the original graph. If you don't believe it, you can grab the original curve and drag it away leaving the derivative underneath.

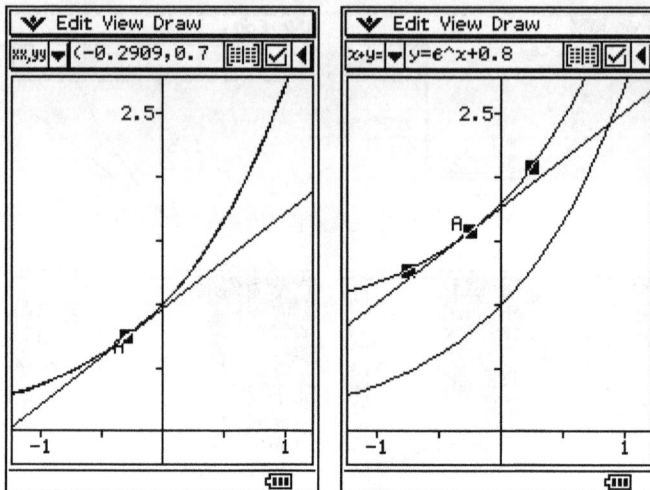

We can confirm this in the eActivity window:

## Derivative of e^x from First Principles

We have seen a graphical illustration that the derivative of the exponential function is the function itself. ClassPad has also verified this for us. We will now attempt to derive this result from first principles (with the help of ClassPad's limit function).

We see that it is possible to factor $e^x$ out of the numerator in the chord slope equation:

▼ File Edit Insert Action

$\dfrac{e^{x+h}-e^x}{h}$

$\dfrac{e^{x+h}-e^x}{h}$

$\dfrac{e^x(e^h-1)}{h}$

$\dfrac{e^x\cdot(e^h-1)}{h}$

□

Alg   Standard Cplx Rad

We can use the limit function to show that the rest of the expression tends to 1 as h tends to 0, and hence the slope tends to $e^x$.

▼ File Edit Insert Action

$\dfrac{d}{dx}\left(e^x\right)$

$e^x$

$\dfrac{e^{x+h}-e^x}{h}$

$\dfrac{e^{x+h}-e^x}{h}$

$\dfrac{e^x(e^h-1)}{h}$

$\dfrac{e^x\cdot(e^h-1)}{h}$

$\displaystyle\lim_{h\to 0}\left(\dfrac{e^h-1}{h}\right)$

1

Alg   Standard Cplx Rad

EXERCISE

1.  Find the derivative of cos(x) from first principles.

## Slope Arithmetic

We have derived from first principles the derivatives of a few simple functions. Do we have to do this for all possible functions? Fortunately the answer is no. We can build the derivatives for more complicated functions from our knowledge of these simpler functions and a few simple rules. We'll start off looking at the sum of two functions:

Put a Geometry Strip into your eActivity, and create the following two functions:

$$\cos(x)$$

$$\frac{x^2}{4} - 2$$

(You should zoom in on them and turn on the integer grid to aid in accurate placement of points)

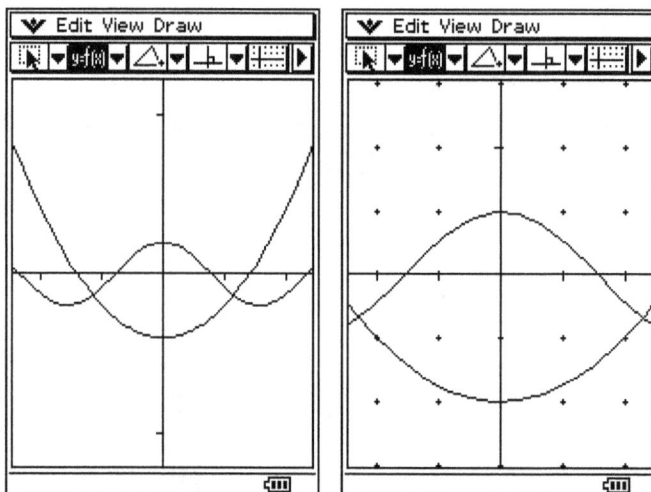

Draw chords on each function from the y intercept to the point at x = 1.  Drag the slope of the chord into the eActivity and add them:

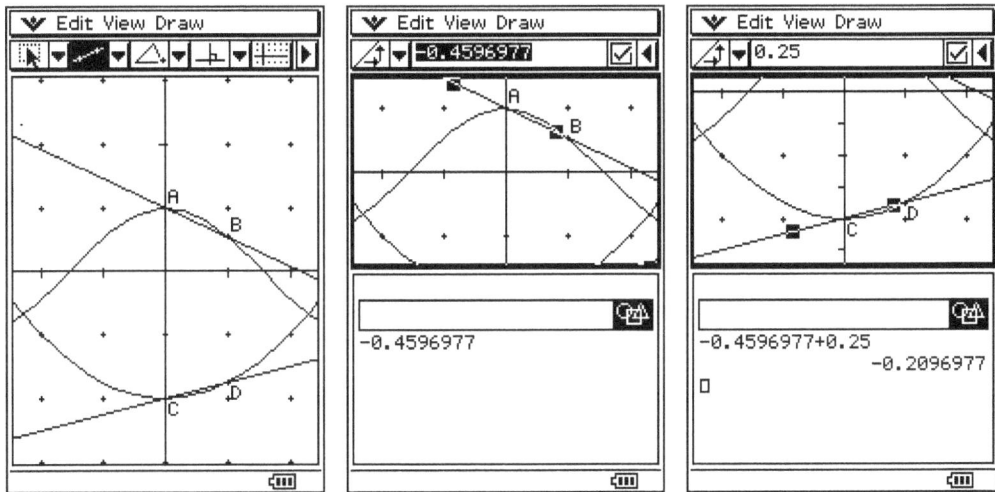

Now, in Geometry, create the function $\cos(x) + \dfrac{x^2}{4} - 2$ (making it thicker can help separate it from the component functions):

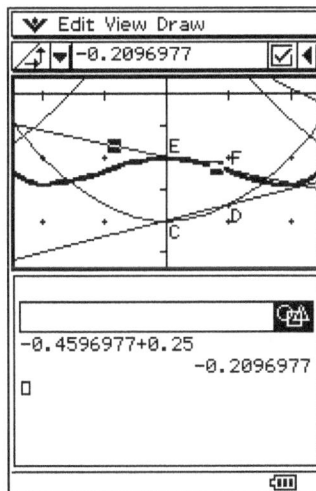

We see that the slope of the sum is the sum of the slopes.

We can check this out algebraically in eActivity: Type in the expression for the slope of the sum. We see we can re-arrange the terms to give the sum of the slopes (some dragging and dropping is necessary):

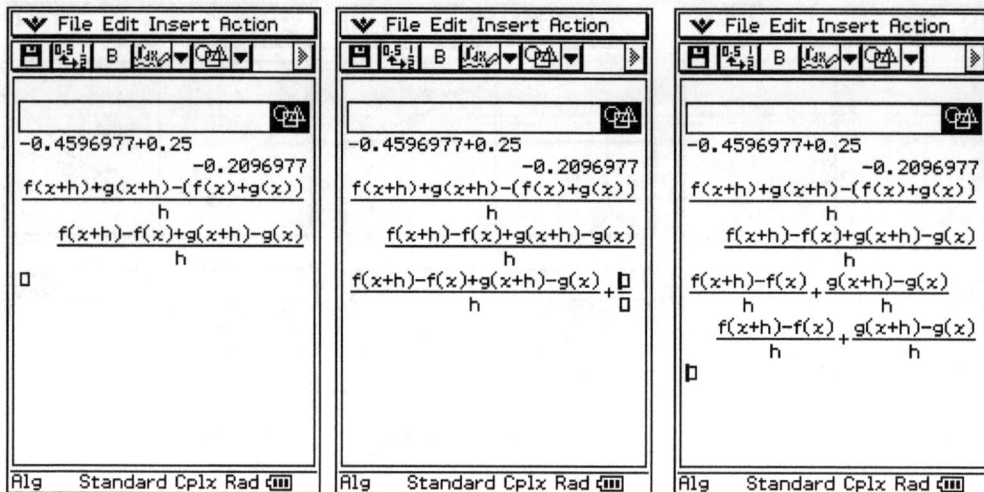

If the chord slope equations add up, so must the tangent slope equations and hence the derivatives. We can check this too in eActivity:

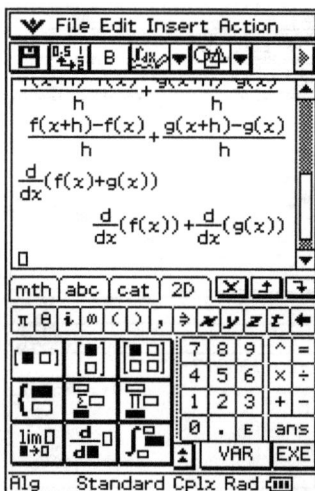

And finally, we can check it out in the geometry side of things: here we are creating tangents at x = -1:

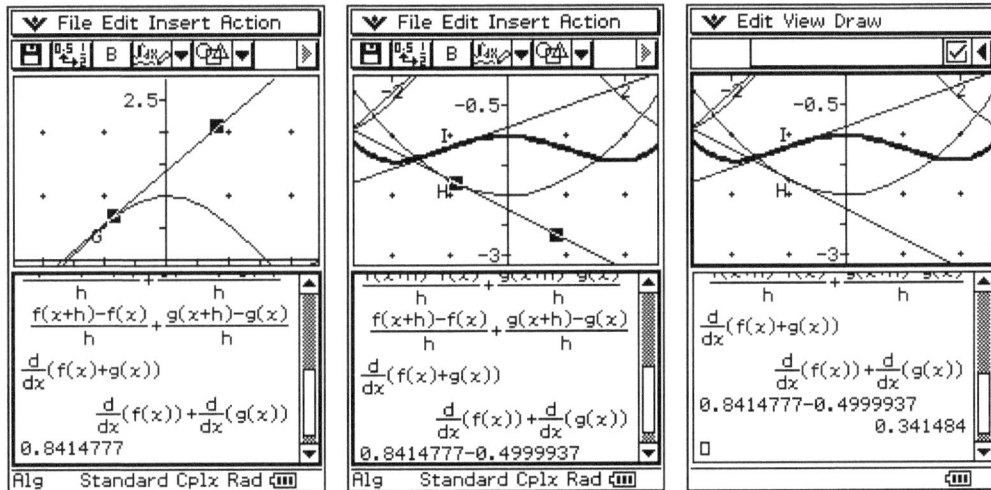

We see that the tangent slopes add up (to within the accuracy of the geometry system).

## Constant Times Derivative

In the previous section, we saw that the derivative of the sum is the sum of the derivatives. So we can now find the derivative of sums of our elementary functions. For example:

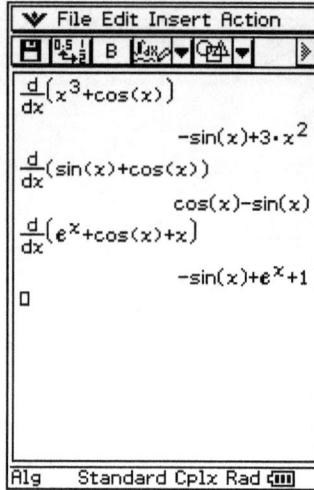

```
▼ File Edit Insert Action

d/dx(x³+cos(x))
                    -sin(x)+3·x²
d/dx(sin(x)+cos(x))
                    cos(x)-sin(x)
d/dx(eˣ+cos(x)+x)
                    -sin(x)+eˣ+1
□

Alg    Standard Cplx Rad
```

How about the derivative of u*f(x), where u is a constant?

First, let's assume u is a whole number, 3 for example. Clearly 3*f(x) = f(x)+f(x)+f(x) and the additive property for derivatives gives us the result that

$$\frac{d(3f)}{dx} = \frac{d(f+f+f)}{dx} = 3\frac{df}{dx}$$

How about a general constant u. Let's examine chords. In a Geometry Strip, create the functions cos(x) and $\dfrac{5\cos(x)}{2}$. You should display the integer grid to help with the location of chord end points, and zoom in appropriately

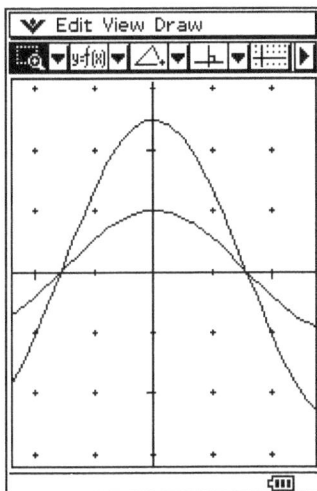

Create a chord from x=0 to x=1 on the cos(x) curve, and another on the $\dfrac{5\cos(x)}{2}$ curve.

Verify in eActivity that the ratio of the chord slopes is $\dfrac{5}{2}$

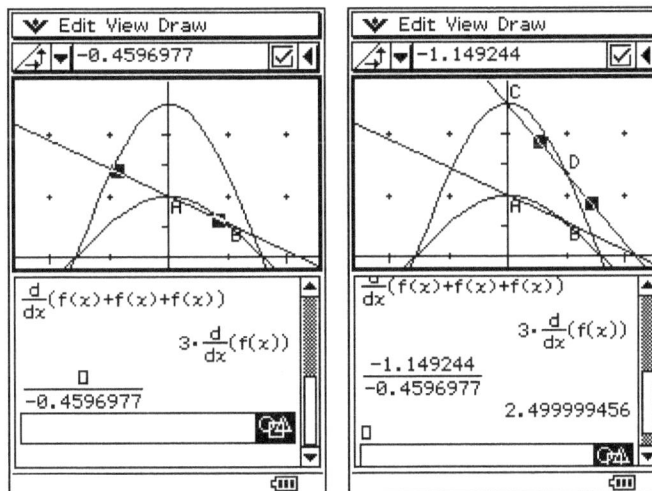

Let's look at the situation algebraically. We can enter the formula for he slope of the chord to the function u.f(x). Applying factor() pulls the constant u out:

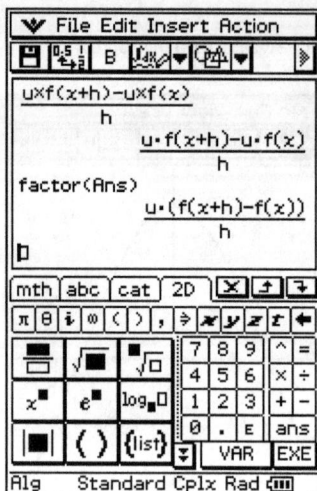

```
▼ File Edit Insert Action
┌─────────────────────────────────┐
│ u×f(x+h)-u×f(x)                  │
│ ─────────────                    │
│       h                          │
│              u·f(x+h)-u·f(x)     │
│              ───────────────     │
│                    h             │
│ factor(Ans)                      │
│              u·(f(x+h)-f(x))     │
│              ───────────────     │
│                    h             │
│ □                                │
└─────────────────────────────────┘
 mth  abc  cat  2D  [X][↲][↳]
 π θ i ω ( ) , ⇒ 𝒙 𝒚 𝒛 𝒕 ←
 ▤  √■  ⁿ√□      7 8 9 ^ =
                 4 5 6 × ÷
 x■  e■  log■□   1 2 3 + -
                 0 . E ans
 ▣  ( )  {list}  ⇟ VAR EXE
 Alg    Standard Cplx Rad ▭
```

Of course, we could have used ClassPad's derivative function in the first place, but that would have been too easy:

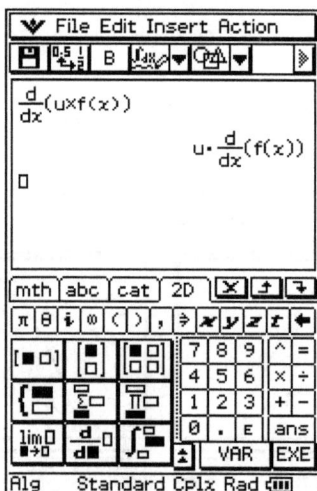

```
▼ File Edit Insert Action
┌─────────────────────────────────┐
│ d                                │
│ ──(u×f(x))                       │
│ dx                               │
│                  d               │
│              u·──(f(x))          │
│                 dx               │
│ □                                │
│                                  │
│                                  │
└─────────────────────────────────┘
 mth  abc  cat  2D  [X][↲][↳]
 π θ i ω ( ) , ⇒ 𝒙 𝒚 𝒛 𝒕 ←
 [■□]  [■□]  [■□]  7 8 9 ^ =
                   4 5 6 × ÷
 {■□  Σ□  ∏□      1 2 3 + -
                   0 . E ans
 lim□ d□ ∫□□      ⇕ VAR EXE
 Alg    Standard Cplx Rad ▭
```

EXERCISES

Write down the derivatives of the following functions.  Check the answers with ClassPad.

1.  $3 \cdot \sin(x)$

2.  $2 \cdot \cos(x) - 5 \cdot \sin(x)$

3.  $\dfrac{e^x}{2}$

## A Special Family of Polynomials

Now we can differentiate the sum of two functions and also the product of a function times a constant.  So, for example, we can look at:

```
▼ File Edit Insert Action
```

$\frac{d}{dx}(2\times\cos(x)+3\times\sin(x))$

$3\cdot\cos(x)-2\cdot\sin(x)$

$\frac{d}{dx}\left(5\times e^x-3\times x^2\right)$

$-6\cdot x+5\cdot e^x$

$\frac{d}{dx}\left(\frac{x^3}{6}+\frac{x^2}{2}+x+1\right)$

$\frac{x^2+2\cdot x+2}{2}$

expand(Ans)

$\frac{x^2}{2}+x+1$

```
Alg    Standard Cplx Rad
```

The last example is interesting.  Let's re-write it in a general form using the summation operator.  First we store the value 3 in the variable n, then define the summation using i! (factorial operator) in the denominator.   We then take the derivative of the expression (drag and drop into place) and do an expand() to get the result into a good form.

```
▼ File Edit Insert Action
```

3⇒n

3

$\sum_{i=0}^{n}\left(\frac{x^i}{i!}\right)$

$\frac{x^3}{6}+\frac{x^2}{2}+x+1$

```
Alg    Standard Cplx Rad
```

```
▼ File Edit Insert Action
```

3⇒n

3

$\sum_{i=0}^{n}\left(\frac{x^i}{i!}\right)$

$\frac{x^3}{6}+\frac{x^2}{2}+x+1$

expand($\frac{d}{dx}\left(\sum_{i=0}^{n}\left(\frac{x^i}{i!}\right)\right)$)

$\frac{x^2}{2}+x+1$

```
Alg    Standard Cplx Rad
```

Now we can look at further polynomials from this family just by increasing the value stored in n:

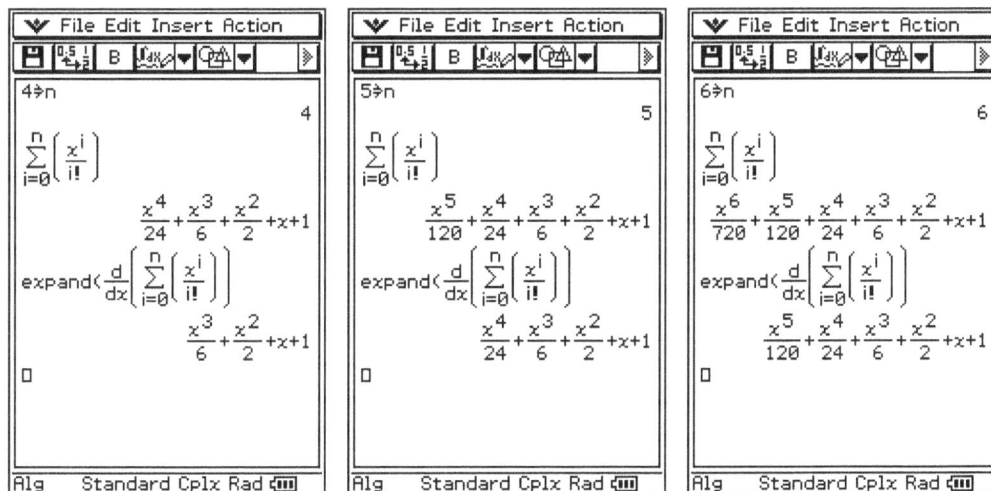

We observe that, in each case, the derivative is the same as the original function without the highest order term. Now if x is suitably small (in particular if $|x|$ is less than 1), we might expect this not to make much difference.

Let's look at that. Drag the 4th order function into a Geometry strip:

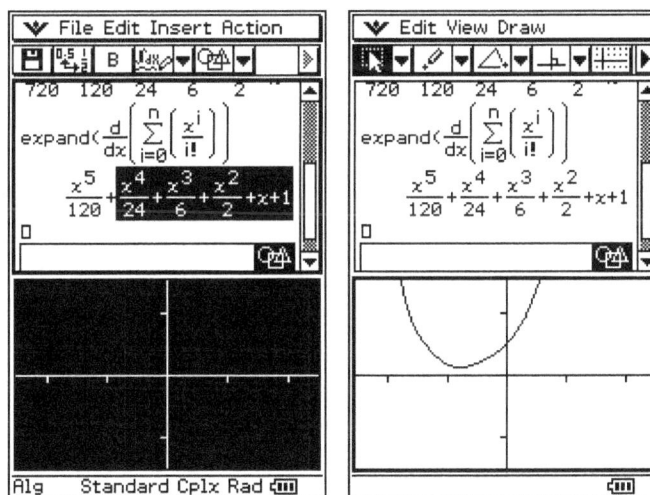

Then drag the 3$^{rd}$ order function

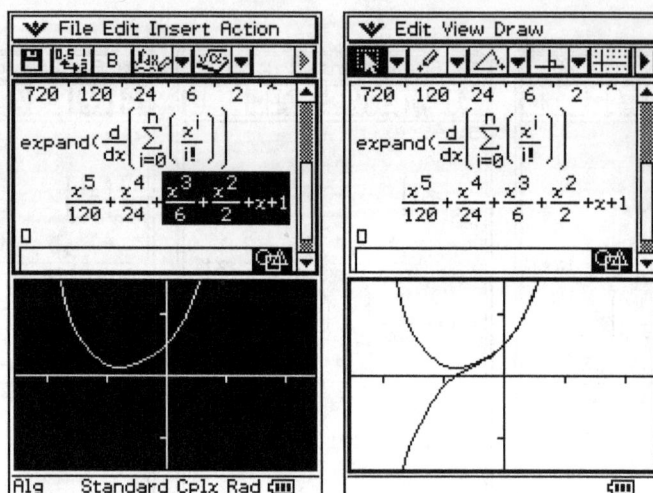

Zooming in appropriately, we see that the two functions resemble each other quite closely in the interval [-1,1]

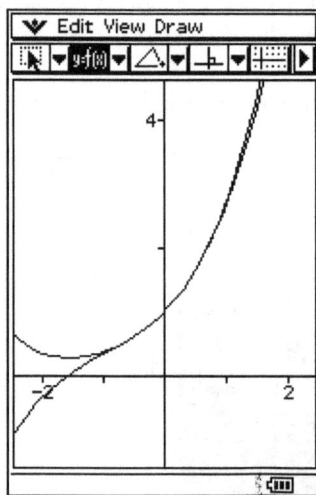

What function do we already know that closely resembles its derivative? In fact resembles it so closely that it is identical?

Try dragging it into the picture:

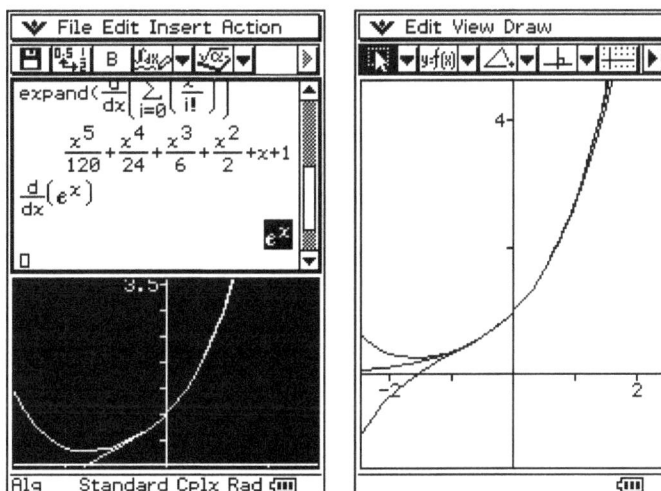

Observe that in the domain [-1,1], $e^x$ is closely mimicked by the polynomial functions.

EXERCISES

1. Evaluate the following expression in ClassPad: $\displaystyle\sum_{i=0}^{3}\frac{(-1)^i x^{2i}}{(2i)!}$

2. Evaluate the following expression in ClassPad: $\displaystyle\sum_{i=0}^{3}\frac{(-1)^i x^{2i+1}}{(2i+1)!}$

3. Differentiate the expression in example 2 and compare with the expression in example 1.

4. Differentiate the expression $\displaystyle\sum_{i=0}^{4}\frac{(-1)^i x^{2i}}{(2i)!}$ and compare with expression 3.

5. Plot the polynomial derived in 1, and compare with a plot of cos(x).

6. Plot the polynomial derived in 2 and compare with a plot of sin(x).

## Derivatives of Products

We know how to differentiate $x^2 + \cos(x)$ and $2.\cos(x)$, but how about $x^2 \cdot \cos(x)$?

Let's try it in ClassPad:

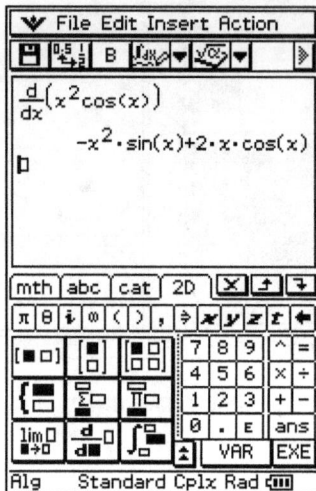

Hmm, perhaps it's not obvious what is going on. Let's look at chord slopes first:

## Mutiplying Chord Slopes

In a Geometry Window, create the functions:

$$2\cos\left(\frac{x}{2}\right)$$

$$\frac{x^2}{4} - 2$$

$$2\cos\left(\frac{x}{2}\right) \cdot \left(\frac{x^2}{4} - 2\right)$$

You should zoom in appropriately, and turn on the integer grid to allow accurate placement of chord end points. (The product function is thicker in the drawing so that it stands out clearly).

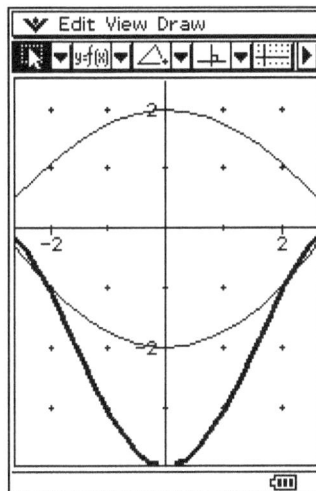

Now create chords from x=0 to x=1 on the two original curves:

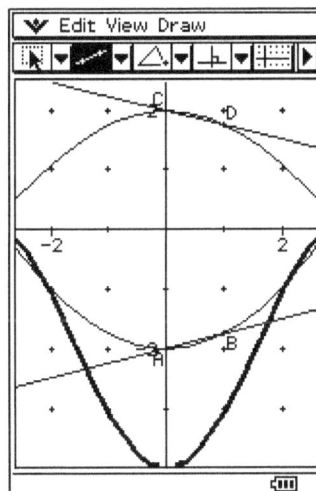

Experience with the sum of the slopes might lead us to hope that the slope of the product is the product of the slopes – if only life were that simple. Let's check that idea out at least:

UNDERSTANDING CALCULUS WITH CLASSPAD

Drag the two slopes into eActivity and multiply them. Check the result against the slope of the chords of the product function:

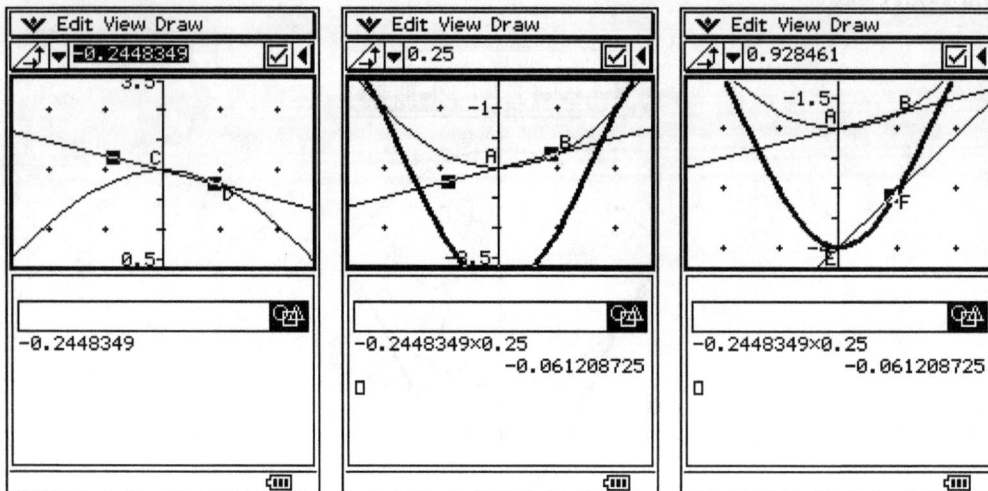

We see that the slope of the resulting chord is not the product of the slopes of the original chords. Not even close.

However, try this. Multiply the slope of the first chord by the y coordinate of the first point on the second chord, and add the slope of the second chord multiplied by the y coordinate of the second point on the first chord.

Eh?

$$slope(AB) \cdot y_C + slope(CD) \cdot y_B$$

Just try it, it's not all that difficult to parse (although hiding the product curve helps make the diagram cleaner)

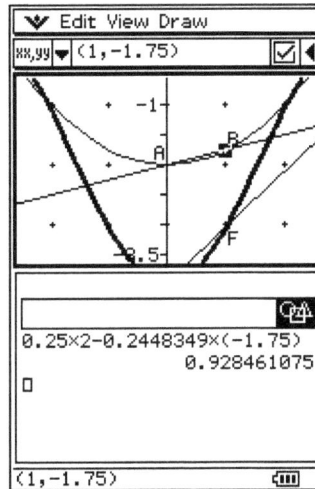

```
┌──────────────────────────────┐
│ ❖ Edit View Draw             │
├──────────────────────────────┤
│ xx,yy ▼ (1,-1.75)     ☑ ◀    │
├──────────────────────────────┤
│           ·  -1·   ·          │
│           ↗         ↗         │
│        ·  ⌣    B̲ ·           │
│        ·    A̲ ·╱  ·          │
│  ·            ╱·              │
│  ·          ╱   ·            │
│         .5·╱  F              │
├──────────────────────────────┤
│  ┌────────────────────┐ ⊕⚖  │
│  │                    │      │
│  0.25×2-0.2448349×(-1.75)    │
│                  0.928461075 │
│  □                           │
│                              │
├──────────────────────────────┤
│ (1,-1.75)              ▭     │
└──────────────────────────────┘
```

Which does in fact correspond to the slope of the chord of the product (to within the accuracy of the geometry system):

We can show that this formula is generally true. First let's define s(x) to be the slope of the chord of f at x, and t(x) to be the slope of the chord of g at x:

```
┌──────────────────────────────┐
│ ❖ File Edit Insert Action    │
├──────────────────────────────┤
│ 🖫 📊 B A✐ ▼ ⊕⚖ ▼        ≫ │
├──────────────────────────────┤
│                f(x+h)-f(x)   │
│ define s(x)=─────────────    │
│                    h         │
│                       done   │
│                g(x+h)-g(x)   │
│ define t(x)=─────────────    │
│                    h         │
│                       done   │
│                              │
│                              │
│                              │
│                              │
│                              │
│                              │
│                              │
│                              │
├──────────────────────────────┤
│ Alg   Standard Cplx Rad ▭   │
└──────────────────────────────┘
```

Now let's evaluate the general expression corresponding to the one we used in the example above:

$$s(x).g(x) + t(x).f(x+h)$$

We'll apply a bit of simplification to the result:

```
▼ File Edit Insert Action
[icons]

define s(x)= f(x+h)-f(x)
              ─────────
                  h
                              done
define t(x)= g(x+h)-g(x)
              ─────────
                  h
                              done
s(x)*g(x)+t(x)*f(x+h)
(f(x+h)-f(x))·g(x)   (g(x+h
─────────────────  + ──────
        h
simplify(Ans)
   g(x+h)·f(x+h)-g(x)·f(x)
   ──────────────────────
              h
□

Alg    Standard Cplx Rad
```

And we see that the formula for the slope of the product function appears like magic.

## Product Rule for Derivatives

The rule for chords should transfer directly to a rule for tangents and hence for derivatives:

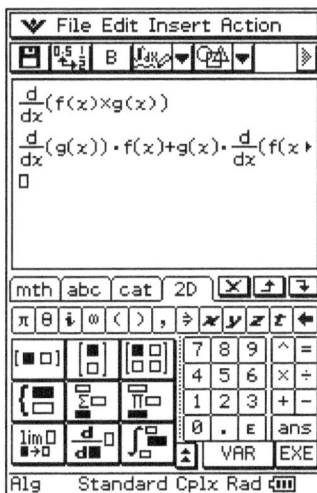

```
W File Edit Insert Action

d
──(f(x)×g(x))
dx
d              d
──(g(x))·f(x)+g(x)·──(f(x ▸
dx             dx
□




mth  abc  cat   2D  [ X ][ ↱ ][ ↴ ]
π θ i ω ⟨ ⟩ , ÷ x y z t ←
[■□]  [■]  [■□]  7 8 9 ^ =
      [□]  [□□]  4 5 6 × ÷
{■   Σ□   π□   1 2 3 + −
 □   □    □    0 . E ans
lim□  d□  ∫□   ↕  VAR  EXE
■→□  d■  ↓□

Alg   Standard Cplx Rad ▥
```

The derivative of f times g is f times the derivative of g plus g times the derivative of f.

Or using f' for the derivative of f:

$$(fg)' = fg' + f'g$$

Now that we can differentiate products, let's differentiate sin(x).cos(x). (A bit of simplification makes the answer look nice:

```
▼ File Edit Insert Action
┌──────────────────────────────┐
│ 🖫 📊 B 📝▼ 📷▼        ≫│
├──────────────────────────────┤
│ define f(x)=sin(x)           │
│                         done │
│ define g(x)=cos(x)           │
│                         done │
│ f(x)×d/dx(g(x))+d/dx(f(x))×g▶│
│        (cos(x))²–(sin(x))²   │
│ simplify(ans)                │
│                   cos(2·x)   │
│ ▯                            │
│                              │
│                              │
│                              │
│                              │
│                              │
├──────────────────────────────┤
│ Alg   Standard Cplx Rad ▩   │
└──────────────────────────────┘
```

Looking at the simple form of the result makes one wonder if the initial equation perhaps does not have a simpler form. Let's try simplifying it:

```
▼ File Edit Insert Action
┌──────────────────────────────┐
│ 🖫 📊 B 📝▼ 📷▼        ≫│
├──────────────────────────────┤
│ define f(x)=sin(x)           │
│                         done │
│ define g(x)=cos(x)           │
│                         done │
│ f(x)×d/dx(g(x))+d/dx(f(x))×g▶│
│        (cos(x))²–(sin(x))²   │
│ simplify(ans)                │
│                   cos(2·x)   │
│ simplify(cos(x)×sin(x)       │
│              sin(x)·cos(x)   │
│ ▯                            │
│                              │
│                              │
│                              │
├──────────────────────────────┤
│ Alg   Standard Cplx Rad ▩   │
└──────────────────────────────┘
```

No good, how about applying the trig collect function tCollect():

```
 W File Edit Insert Action
[icons]                    ▶

define f(x)=sin(x)
                      done
define g(x)=cos(x)
                      done
f(x)×d/dx(g(x))+d/dx(f(x))×g▶
          (cos(x))²-(sin(x))²
simplify(ans)
                   cos(2·x)
tCollect(cos(x)×sin(x)
                   sin(2·x)
                   ───────
                      2
□



Alg    Standard Cplx Rad ⟨▥⟩
```

We see that it has a form f(2.x). We'll learn how to differentiate that sort of function in a later section.

EXERCISES

Write down the following derivatives and check your answer with ClassPad:

1. $x^2 \cos(x)$

2. $e^x \sin(x)$

3. $x \cdot e^x$

4. $\sin(x)^2$

# Derivative of the Reciprocal

If we know the derivative of f(x), can we work out the derivative of $\dfrac{1}{f(x)}$?

As usual, let's look first at the chord slope equation for $\dfrac{1}{f(x)}$.

If we apply the combine() function, we see that the slope of $\dfrac{1}{f(x)}$ is the slope of f(x) divided by the product of the y coordinates of the end points of the chord:

We can check this in an example in geometry.

Draw our favorite function $y = \dfrac{x^2}{4} - 2$ along with the function: $y = \dfrac{1}{\dfrac{x^2}{4} - 2}$.

As usual, zoom in appropriately and view the Integer Grid:

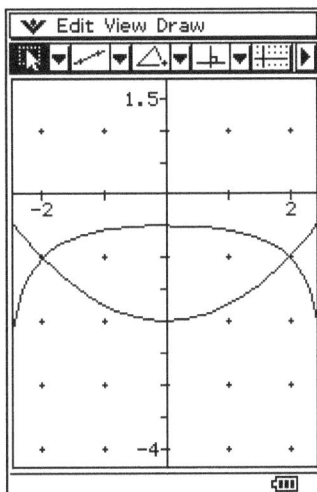

Now create a chord AB through the points with x coordinates 0 and 1 on the first function. Drag the slope of this chord into eActivity and form an expression with the slope as the numerator and the product of the y coordinates of A and B as the denominator. The value of this expression should be the slope of the equivalent chord of the second function. Verify this by creating the chord CD and examining its slope.

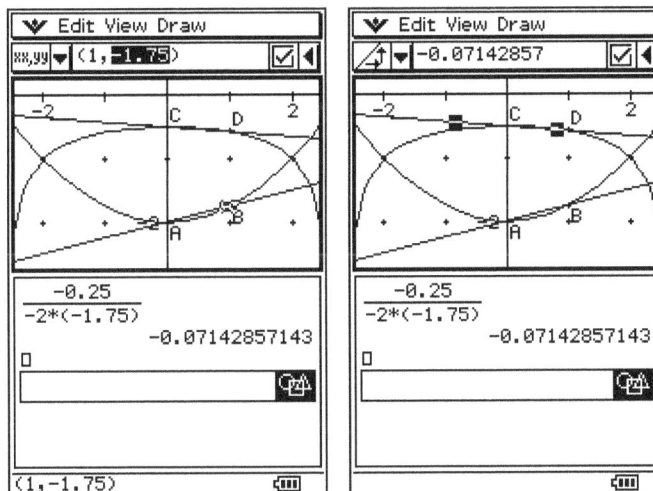

As usual, what is true for slopes of chords is also true for tangents, and thus for derivatives:

```
 W File Edit Insert Action
 ▣ ▣  B  ▨▨▼▨▼         ▶
 d ( 1  )
 dx( f(x) )
                    - d (f(x))
                      dx
                    ───────────
                      (f(x))²
 ▯

 mth abc cat 2D  ▣ ▣ ▣
 π θ i ω ( ) , ⇒ x y z t ←
 [■□]  [■]  [■□]  7 8 9 ^ =
       [□]  [□□]  4 5 6 × ÷
 {■□  Σ□  π□    1 2 3 + -
      □□  □□
 lim□  d □  ∫□    0 . ε ans
 ■→□  d■   □    VAR  EXE
 Alg   Standard Cplx Rad ▥
```

We can use this formula to differentiate sec(x) and csc(x). Check the ClassPad solution to see how it is derived from the formula:

```
 W File Edit Insert Action
 ▣ ▣  B  ▨▨▼▨▼         ▶
 d ( 1    )
 dx( sin(x) )
                    -cos(x)
                    ─────────
                    (sin(x))²
 simplify(ans)
                      -1
                    ──────────────
                    sin(x)·tan(x)
 d ( 1    )
 dx( cos(x) )
                     sin(x)
                    ─────────
                    (cos(x))²
 simplify(ans)
                     tan(x)
                    ─────────
                     cos(x)
 ▯

 Alg   Standard Cplx Rad ▥
```

1.  Apply the formula for derivative of a reciprocal to get the derivative of $\dfrac{1}{x^2}$. Show this is the same as the expression $2x^{-3}$.

2.  Apply the formula for derivative of a reciprocal to get the derivative of $\dfrac{1}{x^n}$. Show this is the same as the expression $-nx^{-n-1}$.

3.  Apply the formula for derivative of a reciprocal to get the derivative of $\dfrac{1}{e^x}$. Show this is the same as the expression $-e^{-x}$.

## Quotients

We can use our formula for the derivative of the reciprocal along with our formula for the derivative of a product to derive the derivative of the quotient of two functions

First we define a function r(x) to be the reciprocal of g(x). then f(x)*r(x) is equivalent to $\frac{f(x)}{g(x)}$.

Applying the product rule to f(x)*r(x) gives a formula for the derivative of the quotient:

Inspection of this equation shows that it looks like this:

$$\frac{f'g - fg'}{g^2}$$

This is frequently referred to as (you guessed it) the **quotient rule**.

Using the quotient rule, we can work out the derivative of tan(x), and cot(x). Check the ClassPad solution to see how it is derived from the formula:

# Chain Rule

We'd now like to extend our differentiation capabilities to functions of the form f(g(x)).

Examples include such functions as:

$$\sin(2x)$$

$$\sin(x^2)$$

$$\sin(x)^2$$

$$e^{\cos(x)}$$

ClassPad can certainly differentiate these functions – can you deduce the rule for the derivative of f(g(x))?

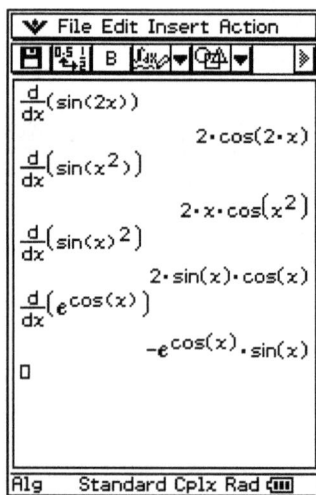

Here are some semi-generic ones which may help you to pin down the rule. (ClassPad will not give the game away by differentiating the entirely generic function)

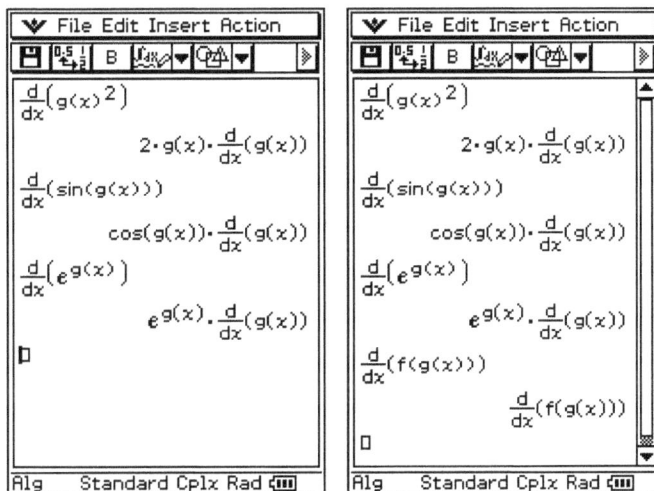

Let's look at chord slope equations. Defining u(x) to be the slope of a chord of the function f(x) defined at the points with x coordinates g(x) and g(x+h) (cunningly contrived to make this all work). Let v(x) be the slope of a chord to the function g(x) defined at the points with x-coordinates x and x+h.

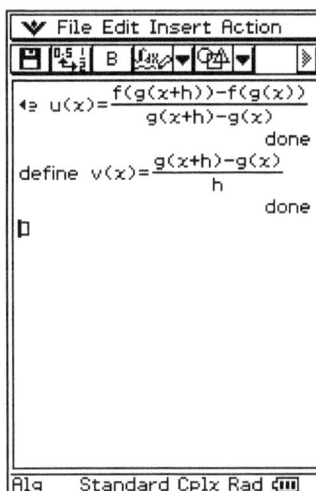

Multiplying these two slopes together, we see that the result is the slope of a chord of the function f(g(x)) through the points with x-coordinates x and x+h.

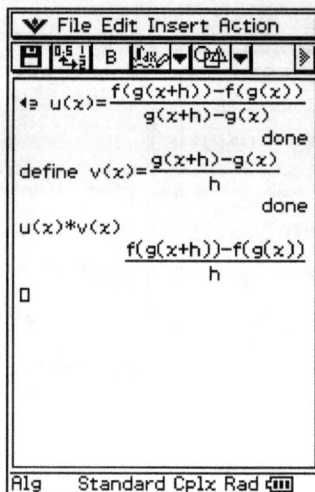

```
╓─  ▼ File Edit Insert Action ─────┐
│ 🖫 │🔢│ B │🔲▼│🔲▼│        ⏩  │
├──────────────────────────────────┤
│            f(g(x+h))-f(g(x))      │
│ ◂🔢 u(x)=──────────────────       │
│            g(x+h)-g(x)            │
│                           done    │
│            g(x+h)-g(x)            │
│ define v(x)=─────────────         │
│                 h                │
│                           done    │
│ u(x)*v(x)                        │
│            f(g(x+h))-f(g(x))      │
│            ─────────────────      │
│                  h               │
│ ▫                                │
│                                  │
│                                  │
│                                  │
├──────────────────────────────────┤
│ Alg    Standard Cplx Rad 🔋      │
└──────────────────────────────────┘
```

## Chain Rule Illustration Using Chords

Let's look at the function:

f(x)=cos(2x)

In a Geometry Window we create the function cos(2x)

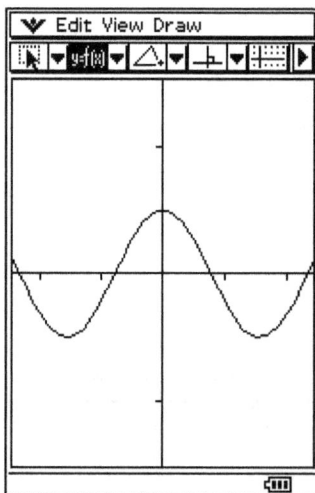

```
╓─  ▼ Edit View Draw ──────────┐
│ 🔲▼│📊▼│△▼│⊥▼│▦│▶         │
├──────────────────────────────┤
│               │              │
│               │              │
│          ╱‾‾╲ │              │
│   ╲     ╱    ╲│   ╱‾‾╲       │
│────╲───╱─────┼──╱────╲───    │
│     ╲_╱      │ ╱      ╲_     │
│               │              │
│                       🔋     │
└──────────────────────────────┘
```

We also create the line y=2x. This should be drawn using the line tool rather than the function tool, as some of the constructions which we will do later in the example work if it is defined as a line, but do not work if it is defined as a function. Use the line tool to sketch the line then enter its equation in the Measure Box.

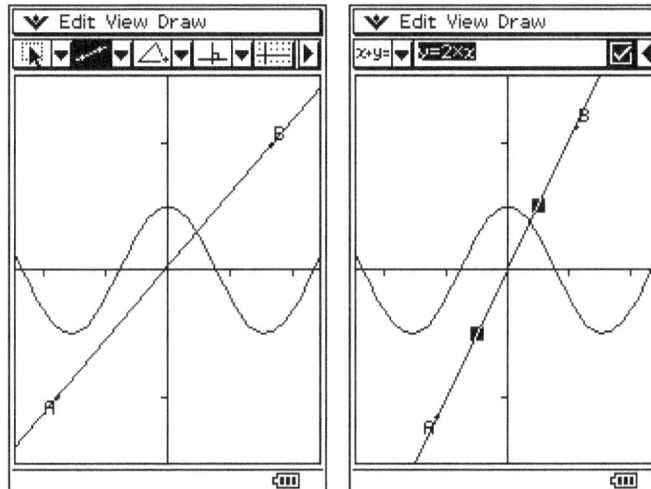

We now wish to place 2 points (defining a chord) on the curve. (Making the points thicker makes them more obvious in a diagram which is about to get cluttered.) We also add the curve y=cos(x)

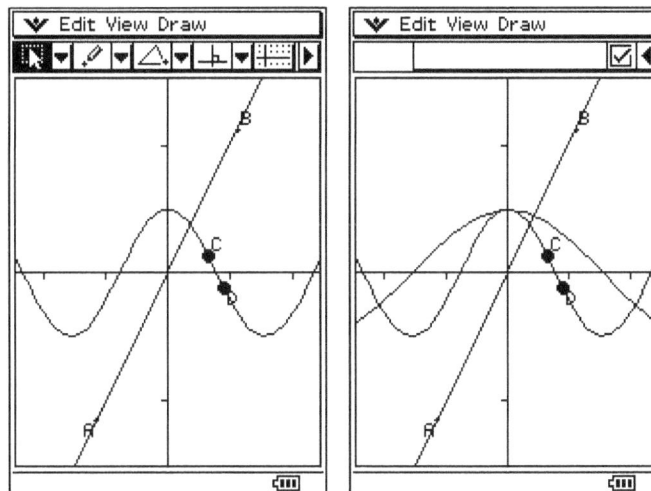

Next we construct lines from C and D to the line AB, and constrain them to be vertical:

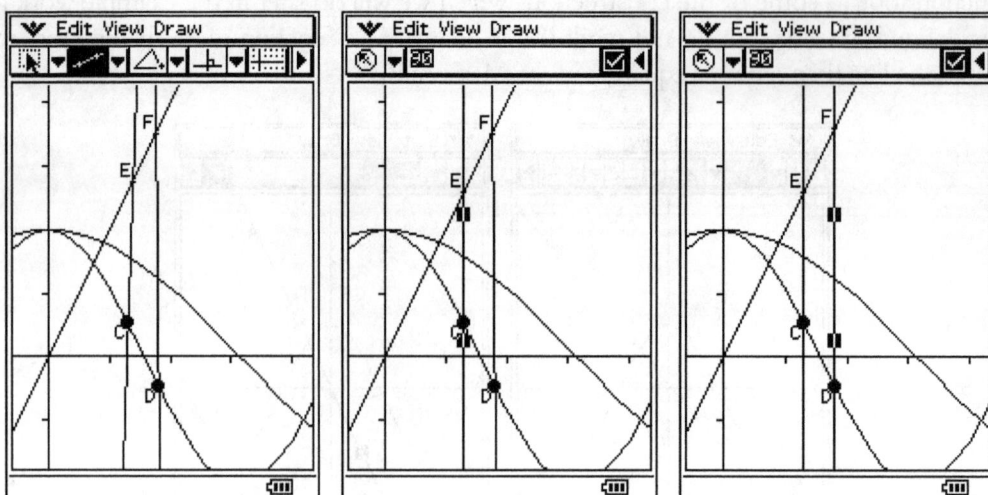

Things are starting to get cluttered, so let's thicken the original curves and the line y=2x and also the new points E and F. While we are at it we can get rid of the axes:

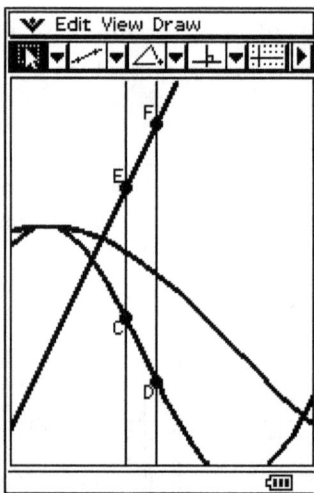

Our next step is to add the line y=x (again use the line tool to sketch the line then set its equation in the Measure Box):

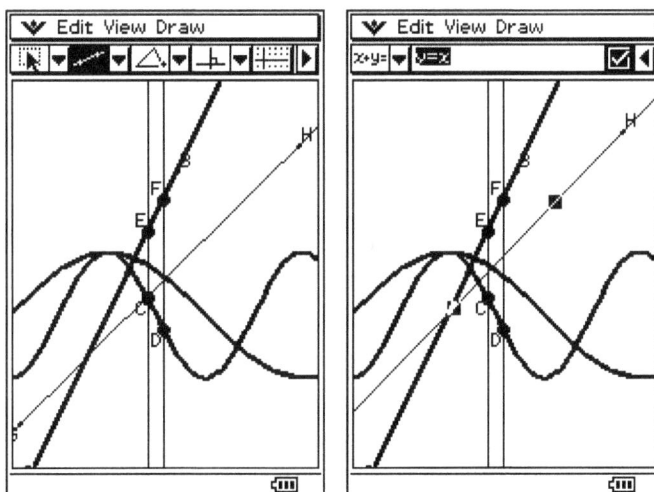

Now, we draw lines from E and F to y=x and constrain them to be horizontal:

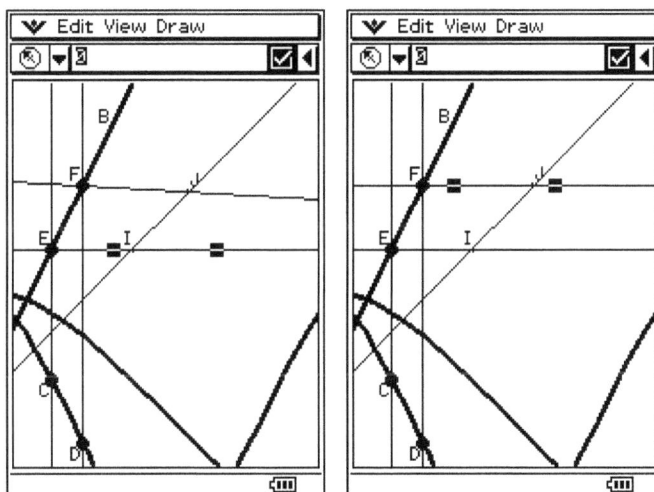

Next, reflect these horizontal lines in y=x:

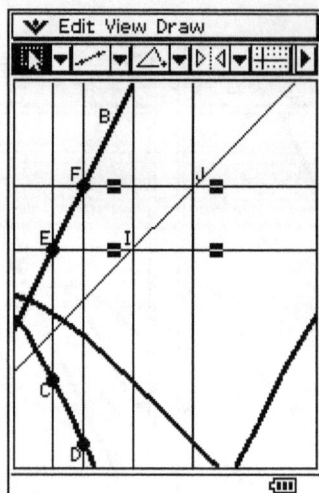

Create a horizontal line from C to the first of these vertical lines (Make sure the new point lies on the line rather than on the curve – to do this you'll have to make the line quite slanted at first, then constrain it to be horizontal.)

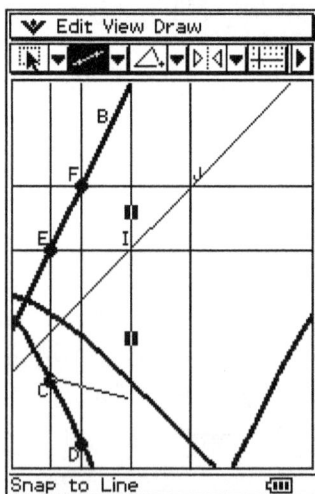

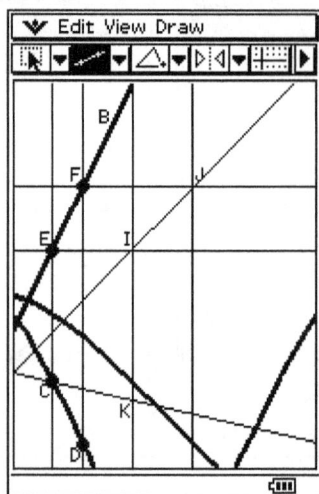

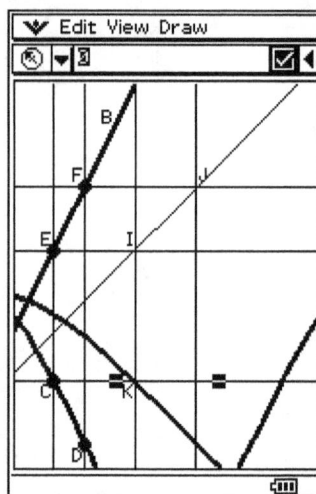

In a similar fashion, create a horizontal line from D to the second vertical line. It is a good idea to thicken the points to enhance their visibility.

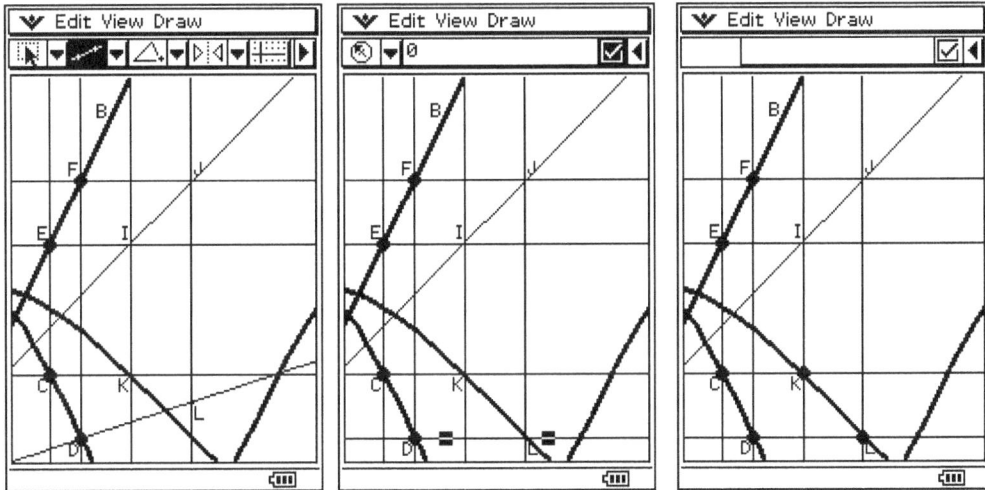

We notice that as C and D traverse the curve

$$y=\cos(2x)$$

K and L traverse the curve

$$y=\cos(x)$$

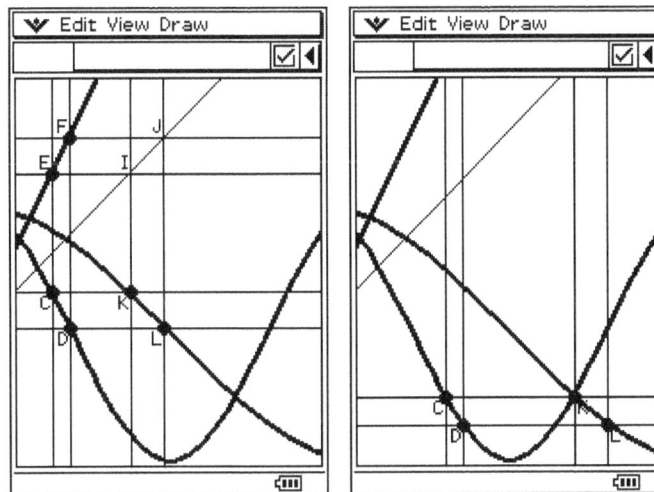

We can see that:

$$slope(CD) = \frac{y_D - y_C}{x_D - x_C} = \frac{y_D - y_C}{x_F - x_E}$$

However

$$slope(EF) = \frac{y_F - y_E}{x_F - x_E}$$

Hence

$$slope(CD) = slope(EF) \cdot \frac{y_D - y_C}{y_F - y_E} = slope(EF) \cdot \frac{y_D - y_C}{x_L - x_K}$$

But K and L are the same height as C and D, so:

$$slope(CD) = slope(EF) \cdot \frac{y_L - y_M}{x_L - x_K} = slope(EF) \cdot slope(KL)$$

## Chain Rule for Derivatives

The slope formula above converts into the following rule for the derivative of a compound function:

$$\frac{df}{dx} = \frac{df}{dg}\frac{dg}{dx}$$

This is called the **Chain Rule**.

In a Geometry Window, create the function y=sin(x), and also a tangent to the function. Bring up the measure box and examine the slope

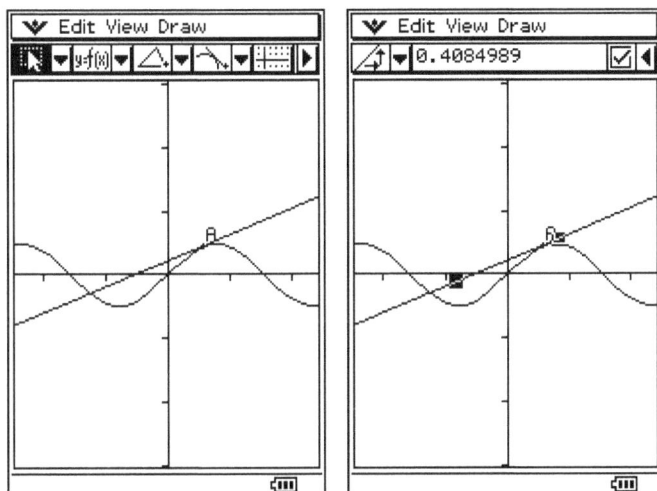

Now create a geometry link in eActivity and drag in the function:

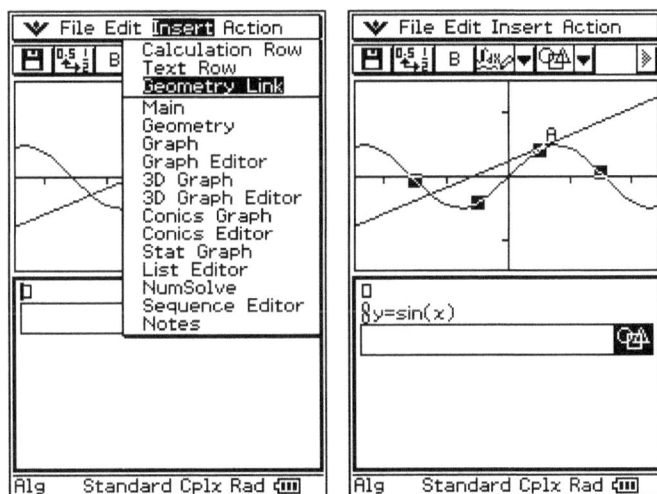

Now select **both** the point A **and** the function and drag them to another location. Observe how the function changes in the Geometry Link. Notice that the slope of the tangent line stays the same:

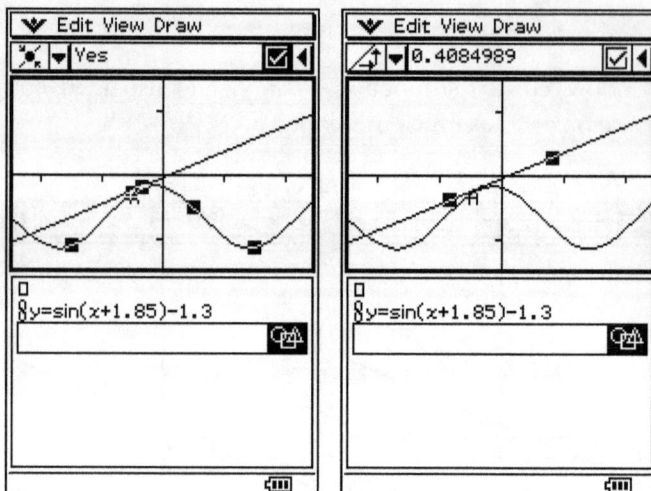

You'll see that dragging the function by a vector (A,B) changes it from

$$y=\sin(x)$$

to

$$y=\sin(x - A)+B.$$

Let's differentiate this:

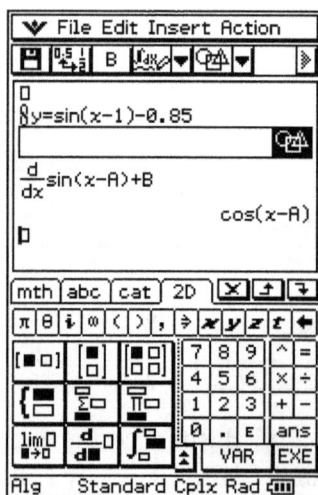

Observe that the slope of the new tangent at the point x+A is the same as the slope of the old tangent at the point x.

By contrast, if you select only the function and drag it, leaving the point A in place, you see that the slope of the tangent changes:

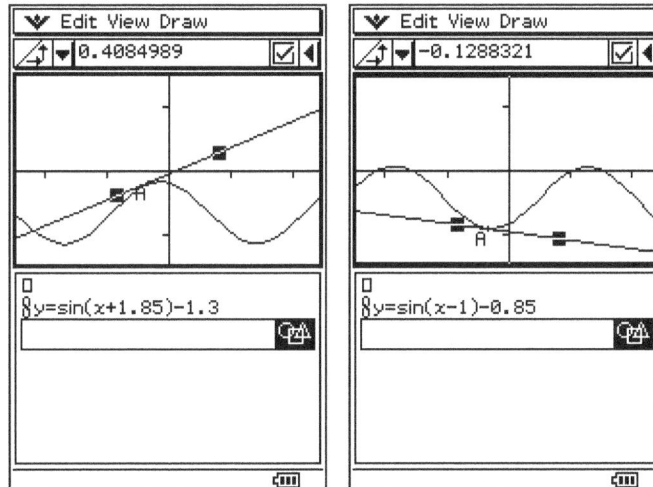

## Chain Rule Application

Let's apply the chain rule to the functions we mentioned at the top of the chapter.

First, we'll look at $\sin(x^2)$. We define functions:

$$f(x) = \sin(x)$$

$$g(x) = x^2$$

Then

$$f(g(x)) = \sin(x^2)$$

We can write the chain rule formula by differentiating f(u) with respect to u and substituting g(x) for u:

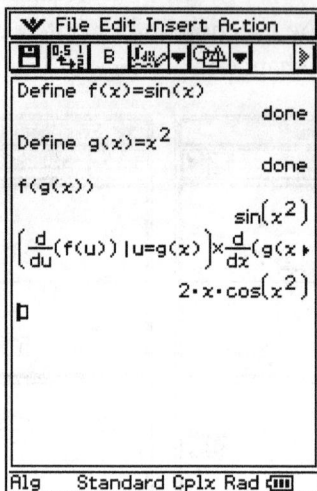

```
▼ File Edit Insert Action
[💾][📊] B [≈▼][📷▼]            [》]
Define f(x)=sin(x)
                              done
Define g(x)=x²
                              done
f(g(x))
                          sin(x²)
(d/du(f(u))|u=g(x))×d/dx(g(x ▸
                       2·x·cos(x²)
▯
Alg    Standard Cplx Rad [▮▮]
```

Having set up our chain rule eActivity, it is easy to apply it to different functions - just change the two Define lines, make sure the cursor is in the <u>first</u> line and hit EXE to recalculate the page:

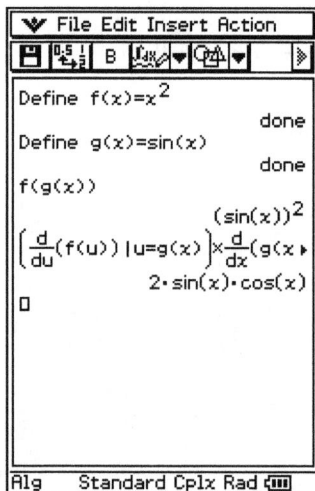

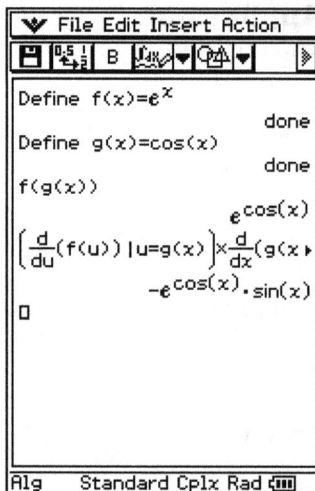

```
▼ File Edit Insert Action
[💾][📊] B [≈▼][📷▼]            [》]
Define f(x)=x²
                            done
Define g(x)=sin(x)
                            done
f(g(x))
                        (sin(x))²
(d/du(f(u))|u=g(x))×d/dx(g(x ▸
                   2·sin(x)·cos(x)
▯
Alg    Standard Cplx Rad [▮▮]
```

```
▼ File Edit Insert Action
[💾][📊] B [≈▼][📷▼]            [》]
Define f(x)=eˣ
                            done
Define g(x)=cos(x)
                            done
f(g(x))
                          e^cos(x)
(d/du(f(u))|u=g(x))×d/dx(g(x ▸
                  -e^cos(x)·sin(x)
▯
Alg    Standard Cplx Rad [▮▮]
```

EXERCISES

Apply the chain rule to evaluate the following derivatives:

1. $\sin(\cos(x))$

2. $e^{x^2}$

3. $\tan(2x)$

4. $e^{\sin(x)}$

# Implicit Differentiation

We're getting close to being able to differentiate any function, but there are a few we can't do yet: for example:

$$f(x) = \sqrt{x}$$

$$f(x) = \ln(x)$$

$$f(x) = \tan^{-1}(x)$$

We'll learn how to do these using a technique called **implicit differentiation**.

Wait a minute, you say, we already know the first one, we can just write it as:

$$f(x) = x^{\frac{1}{2}}$$

And we already know that

$$\frac{d}{dx}x^n = nx^{n-1}$$

Yes, but if you look back to the section where we proved this result, we only proved it for integer values of n.

However, we can use the integer value result to show that the formula is true for rational values of n. And in the process come up with a general technique which can be used to work out the derivative of the inverse functions.

## Derivative of xⁿ for Rational n

First, we use f(x) as a disguise for $\sqrt{x}$ , so that ClassPad does not simplify things out too soon, and write the equation which defines f(x):

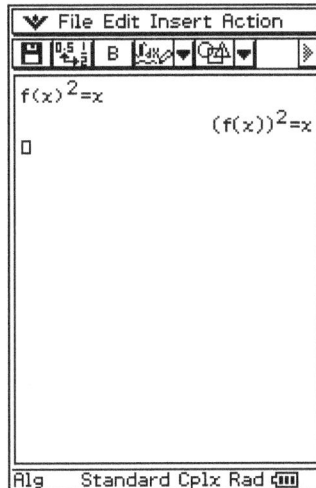

```
┌─────────────────────────────────┐
│ ▼ File Edit Insert Action       │
├─────────────────────────────────┤
│ 🖫 📊 B 📈▼ 📐▼        ▶         │
├─────────────────────────────────┤
│ f(x)²=x                         │
│                   (f(x))²=x     │
│ □                               │
│                                 │
│                                 │
│                                 │
│                                 │
│                                 │
│                                 │
│                                 │
│                                 │
│                                 │
├─────────────────────────────────┤
│ Alg   Standard Cplx Rad ▥       │
└─────────────────────────────────┘
```

We can now differentiate this equation (the right side differentiates to 1):

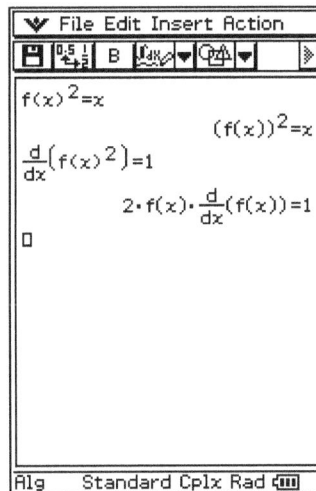

```
┌─────────────────────────────────┐
│ ▼ File Edit Insert Action       │
├─────────────────────────────────┤
│ 🖫 📊 B 📈▼ 📐▼        ▶         │
├─────────────────────────────────┤
│ f(x)²=x                         │
│                   (f(x))²=x     │
│ d                               │
│ ──(f(x)²)=1                     │
│ dx                              │
│                        d        │
│            2·f(x)·──(f(x))=1    │
│                        dx       │
│ □                               │
│                                 │
│                                 │
│                                 │
├─────────────────────────────────┤
│ Alg   Standard Cplx Rad ▥       │
└─────────────────────────────────┘
```

ClassPad has used the Chain Rule to differentiate the left side of the equation. Now, we can solve this equation for $\dfrac{df}{dx}$. ClassPad's solve() command requires you to solve for a variable, so in order to use it, you will have to substitute a variable for the derivative.

Just drag the result into the next input line and replace the derivative with a variable name u, then solve for u:

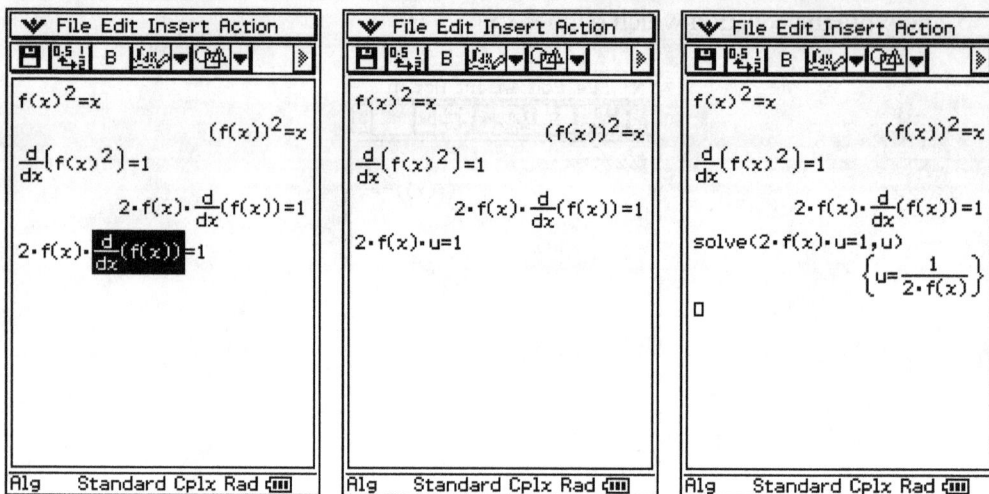

This gives us an expression for the derivative, but it is written in terms of f(x). However, f(x) is a disguise for $\sqrt{x}$, so we can substitute it back in. Again drag the result of the previous computation into an input line and then do the replacement:

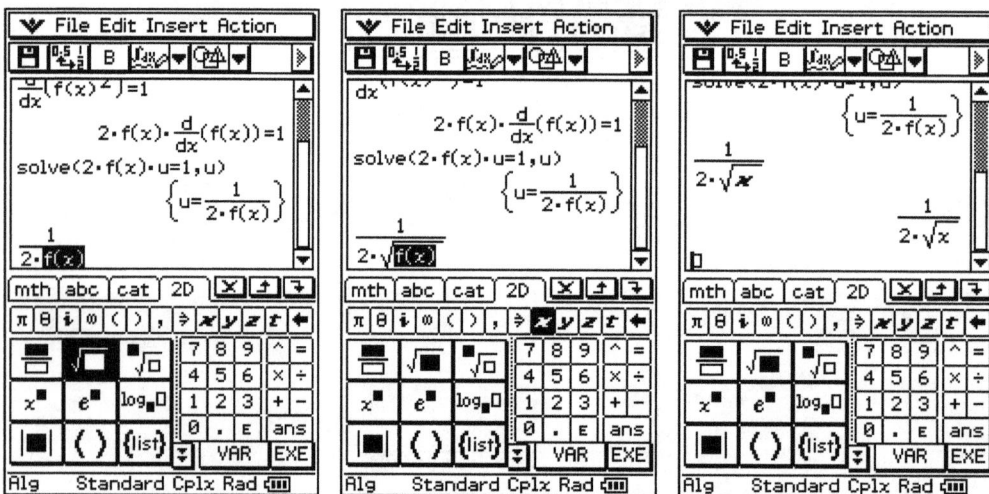

This indeed corresponds to the form $nx^{n-1}$

We'd like now to do similar analyses for other implicitly defined functions. We can restructure our eActivity so that it uses the with command ( denoted by | , which is found on the MATH/OPTN keypad) instead of dragging and dropping and editing in place. This will save effort when we want to look at a new function.

We'll start with the function $f(x) = \sqrt[3]{x}$

To re-run this analysis for a different root, we simply need to change the power in the first line and the root in the third line - remember to hit the EXE key with the cursor on the <u>first</u> line:

| | | |
|---|---|---|
| **File Edit Insert Action** | **File Edit Insert Action** | **File Edit Insert Action** |
| $\frac{d}{dx}\left(f(x)^4\right)=1\mid\frac{d}{dx}(f(x))=u$ | $\frac{d}{dx}\left(f(x)^5\right)=1\mid\frac{d}{dx}(f(x))=u$ | $\frac{d}{dx}\left(f(x)^n\right)=1\mid\frac{d}{dx}(f(x))=u$ |
| $4\cdot u\cdot(f(x))^3=1$ | $5\cdot u\cdot(f(x))^4=1$ | $n\cdot u\cdot(f(x))^{n-1}=1$ |
| solve(ans,u) | solve(ans,u) | solve(ans,u) |
| $\left\{u=\frac{1}{4\cdot(f(x))^3}\right\}$ | $\left\{u=\frac{1}{5\cdot(f(x))^4}\right\}$ | $\left\{u=\frac{(f(x))^{-n+1}}{n}\right\}$ |
| ans$\mid$f(x)=$\sqrt[4]{x}$ | ans$\mid$f(x)=$\sqrt[5]{x}$ | ans$\mid$f(x)=$\sqrt[n]{x}$ |
| $\left\{u=\frac{1}{4\cdot x^{\frac{3}{4}}}\right\}$ | $\left\{u=\frac{1}{5\cdot x^{\frac{4}{5}}}\right\}$ | $\left\{u=\frac{\left(x^{\frac{1}{n}}\right)^{-n+1}}{n}\right\}$ |
| Alg  Standard Cplx Rad | Alg  Standard Cplx Rad | Alg  Standard Cplx Rad |

## Derivative of Inverse Functions

We can use the same type of analysis for other inverse functions. For example, ln(x) is defined to be the inverse of $e^x$. A simple change to the equation in the first line and the definition of f(x) in the last line does the trick:

**File Edit Insert Action**

$\frac{d}{dx}\left(e^{f(x)}\right)=1\mid\frac{d}{dx}(f(x))=u$

$e^{f(x)}\cdot u=1$

solve(ans,u)

$\{u=e^{-f(x)}\}$

ans$\mid$f(x)=ln(x)

$\left\{u=\frac{1}{x}\right\}$

Alg  Standard Cplx Rad

How about those pesky inverse trig functions? No problem with our automated approach to implicit differentiation:

| ▼ File Edit Insert Action |
|---|
| $\frac{d}{dx}(\cos(f(x)))=1\mid\frac{d}{dx}(f(x))\blacktriangleright$ |
| $-u\cdot\sin(f(x))=1$ |
| solve(ans,u) |
| $\left\{u=\frac{-1}{\sin(f(x))}\right\}$ |
| ans$\mid$f(x)=$\cos^{-1}(x)$ |
| $\left\{u=\frac{-1}{\sqrt{-x^2+1}}\right\}$ |
| ☐ |
| Alg    Standard Cplx Rad |

| ▼ File Edit Insert Action |
|---|
| $\blacktriangleleft(\cos(f(x)))=1\mid\frac{d}{dx}(f(x))=u$ |
| $-u\cdot\sin(f(x))=1$ |
| solve(ans,u) |
| $\left\{u=\frac{-1}{\sin(f(x))}\right\}$ |
| ans$\mid$f(x)=$\cos^{-1}(x)$ |
| $\left\{u=\frac{-1}{\sqrt{-x^2+1}}\right\}$ |
| ☐ |
| Alg    Standard Cplx Rad |

| ▼ File Edit Insert Action |
|---|
| $\frac{d}{dx}(\tan(f(x)))=1\mid\frac{d}{dx}(f(x))\blacktriangleright$ |
| $u\cdot\left((\tan(f(x)))^2+1\right)=1$ |
| solve(ans,u) |
| $\left\{u=\frac{1}{(\tan(f(x)))^2+1}\right\}$ |
| ans$\mid$f(x)=$\tan^{-1}(x)$ |
| $\left\{u=\frac{1}{x^2+1}\right\}$ |
| ☐ |
| Alg    Standard Cplx Rad |

| ▼ File Edit Insert Action |
|---|
| $\blacktriangleleft\cdot(\tan(f(x)))=1\mid\frac{d}{dx}(f(x))=u$ |
| $u\cdot\left((\tan(f(x)))^2+1\right)=1$ |
| solve(ans,u) |
| $\left\{u=\frac{1}{(\tan(f(x)))^2+1}\right\}$ |
| ans$\mid$f(x)=$\tan^{-1}(x)$ |
| $\left\{u=\frac{1}{x^2+1}\right\}$ |
| ☐ |
| Alg    Standard Cplx Rad |

EXERCISES

1. Use ClassPad to find the derivatives of the following functions:

   a. cosh(x)

   b. sinh(x)

   c. tanh(x)

2. Use implicit differentiation to compute the derivatives of:

   a. $\cosh^{-1}(x)$

   b. $\sinh^{-1}(x)$

   c. $\tanh^{-1}(x)$

3. $(f(x) - A)^2 = B$, where A and B are constants. Write an equation for $\dfrac{df}{dx}$.

## Inverse Functions

In the previous section, we saw how to use the method of implicit differentiation to find the derivative of an inverse function. Here we derive a general rule for inverse functions. First, an example:

In a Geometry Window create the function

$$y(x) = \frac{x^2}{4} - 1$$

and its inverse

$$y^{-1}(x) = \sqrt{4(x+1)}$$

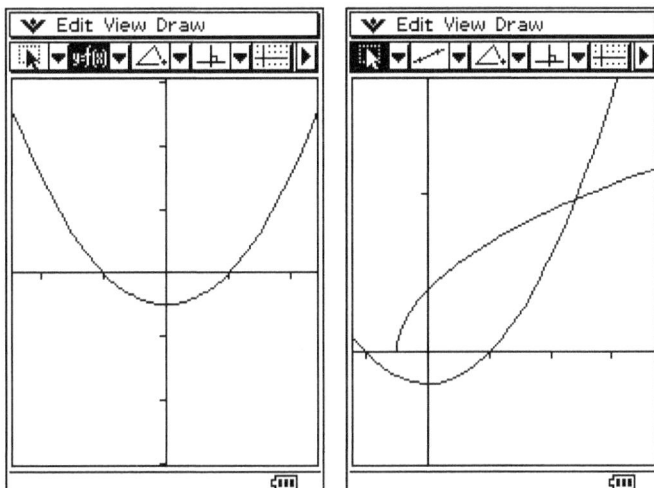

Make a tangent on the original curve. Use **Edit/Properties/Thicker** to make the tangent point more visible.

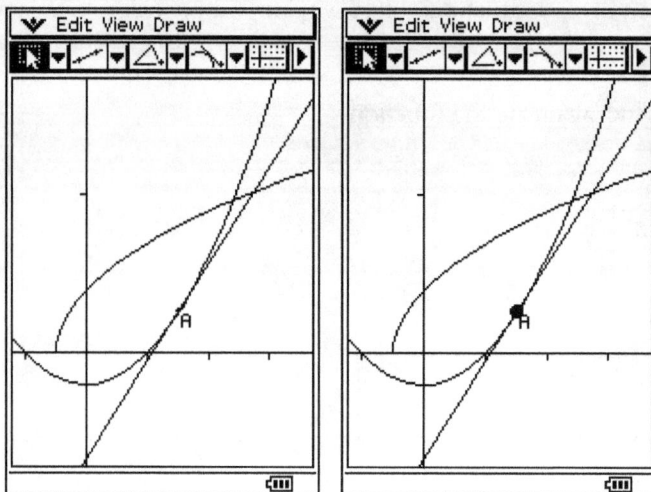

We want to reflect the point in the line y=x. First we need to draw the line – use the line tool rather than the function tool, as ClassPad lets you reflect in a line but not in a function. After drawing the line, set its equation in the Measure Box.

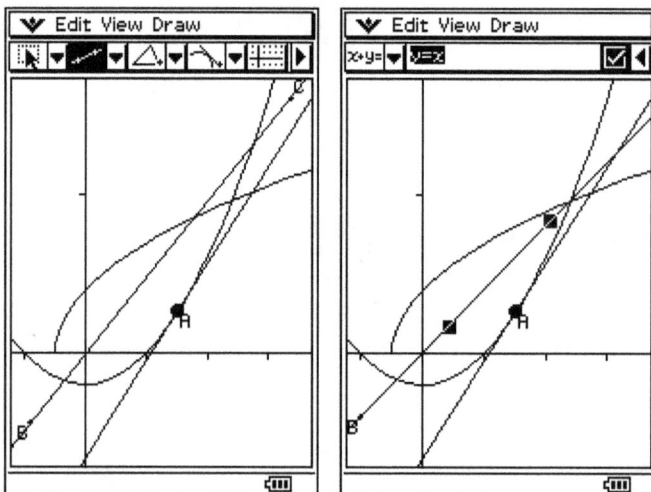

Now reflect point A and the tangent line in the line y=x. As you drag A around y(x), observe its reflection A' move round the inverse function. Also observe that the reflected tangent is the tangent of the inverse curve.

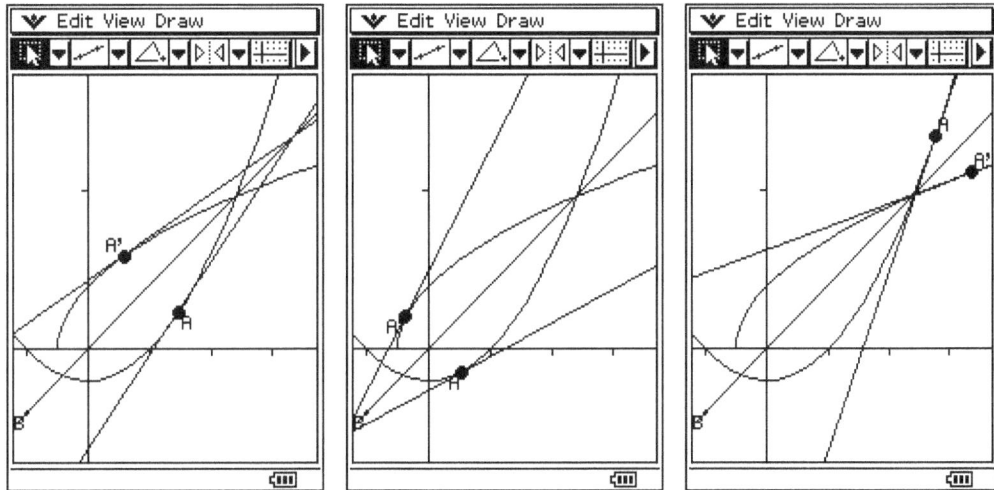

The slope of the reflected line is the reciprocal of the slope of the original tangent, as we can illustrate by adding a triangle to the picture and reflecting it.

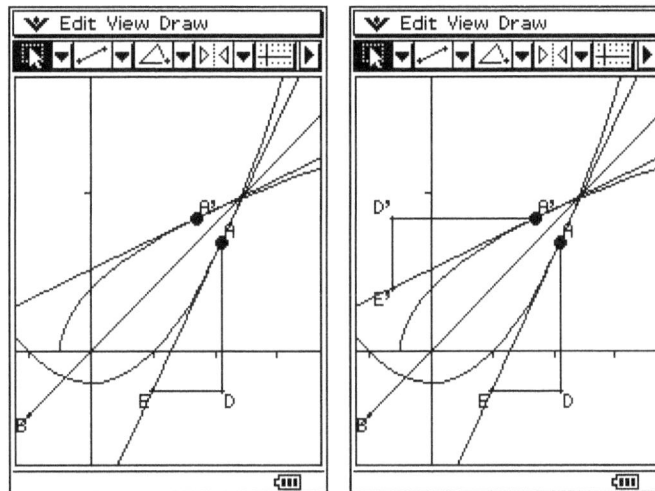

The slope of the first tangent is

$$\frac{|AD|}{|DE|}$$

The slope of the reflected tangent is

$$\frac{|D'E'|}{|A'D'|} = \frac{|DE|}{|AD|}.$$

We can see this from the chain rule.  Let f and g be inverse functions.  Then:

$$f(g(x)) = x$$

Hence

$$\frac{df}{dg}\frac{dg}{dx} = 1$$

And

$$\frac{df}{dg} = \left(\frac{dg}{dx}\right)^{-1}$$

We can apply this formula to find the derivatives of some inverse functions:

| | | |
|---|---|---|
| ▼ File Edit Insert Action | ▼ File Edit Insert Action | ▼ File Edit Insert Action |
| $\frac{d}{dy}(e^y)$ | $\frac{d}{dy}(y^2)$ | $\frac{d}{dy}(\sin(y))$ |
| $e^y$ | $2\cdot y$ | $\cos(y)$ |
| $\frac{1}{ans}$ | $\frac{1}{ans}$ | $\frac{1}{ans}$ |
| $e^{-y}$ | $\frac{1}{2\cdot y}$ | $\frac{1}{\cos(y)}$ |
| ans\|y=ln(x) | ans\|y=√x | ans\|y=sin⁻¹(x) |
| $\frac{1}{x}$ | $\frac{1}{2\cdot\sqrt{x}}$ | $\frac{1}{\sqrt{-x^2+1}}$ |
| □ | □ | □ |
| Alg  Standard Cplx Rad | Alg  Standard Cplx Rad | Alg  Standard Cplx Rad |

Use the above analysis to calculate derivatives for:

1. $\cos^{-1}(x)$

2. $\sinh^{-1}(x)$

## Applications of Differentiation

We've spent a significant amount of effort working out how to calculate the slope of the tangent to a curve. What is this information good for? Was it worth the work?

In this section we'll give a few examples of the applications of differentiation. Not enough, on their own, to convince you that calculus is worth while, but perhaps sufficient to persuade you that there are enough examples in the broad world to justify the trouble of learning the subject.

# Maxima and Minima

Draw this function in a Geometry Window:

$$\frac{x^3}{9} + \frac{x^2}{6} - 2x - 1$$

Then create a tangent. For clarity, you can use **Edit/Properties/Thicker** to make the point stand out:

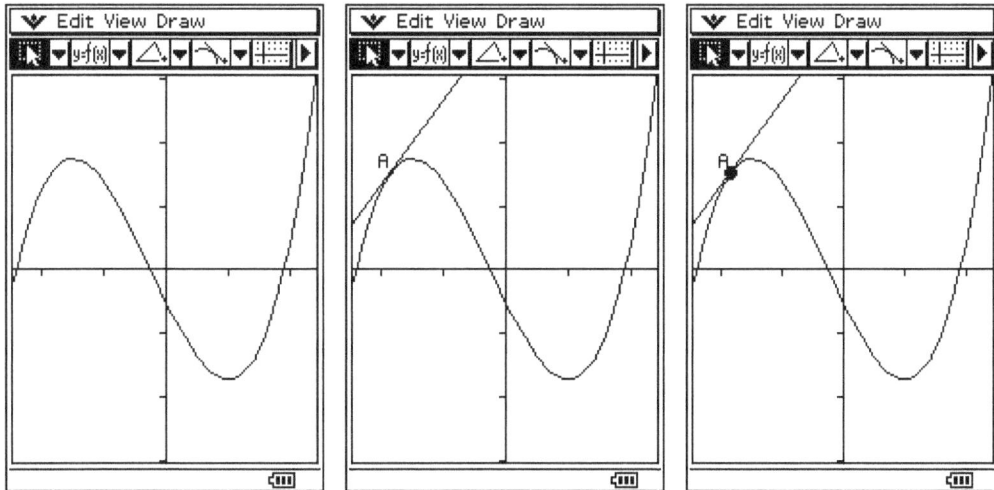

Drag A to the local maximum (the top of the hill that it is sitting on). What do you notice about the tangent?

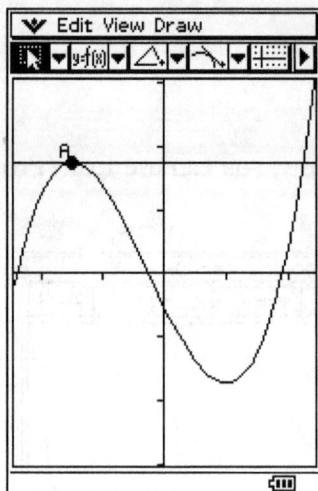

Now drag A until it sits at the bottom of the next trough:

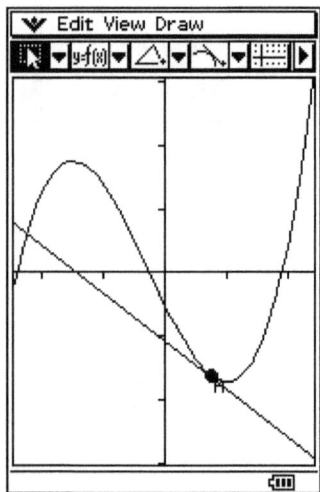

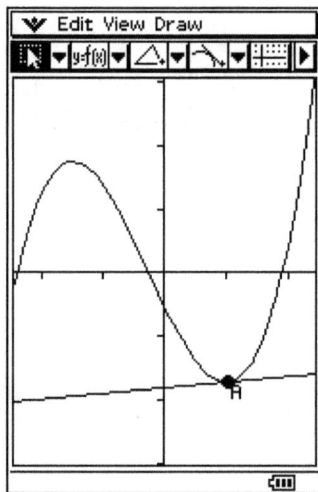

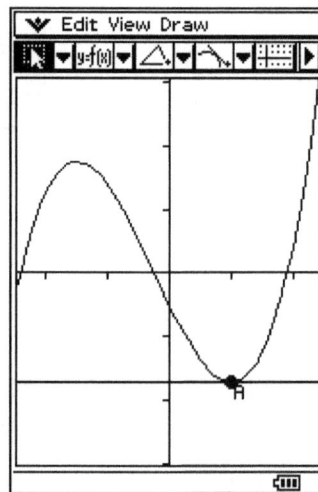

What is the slope of the tangent, when A is at a local maximum, or a local minimum?

Now the slope of the tangent is the derivative of the original function, let's create the derivative curve.

First drag the original curve into eActivity and then differentiate it.

Then drag the resulting curve back into Geometry:

With the derivative safely captured, draw a line through A, then set it's direction so that it is vertical. (You should **not** put B, the other point of the line, on the derivative curve, ClassPad won't cooperate and make the line vertical if you do!)

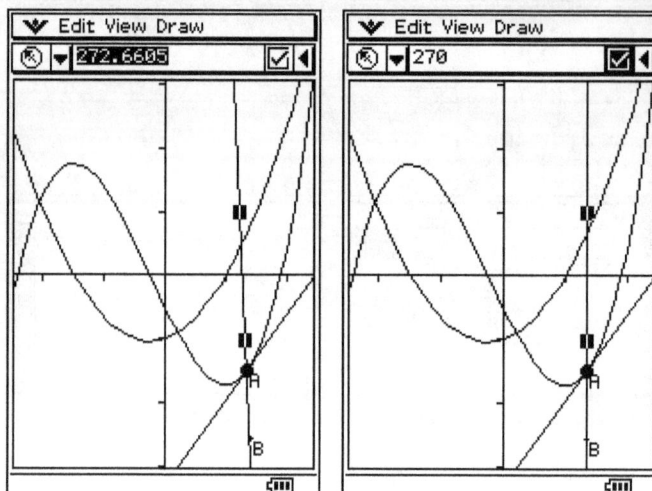

Now drag A until it lies approximately at the minimum.

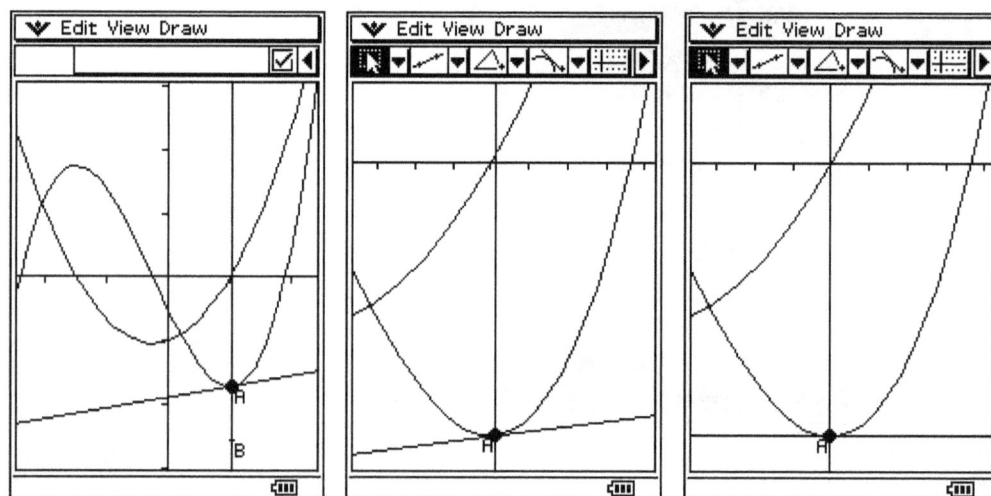

You can see that the x-value where A is at a minimum is the x-value where the derivative curve crosses the x-axis.

We can set up out eActivity to solve this sort of problem. First, we define f(x) to be the function whose maximum or minimum we are interested in. Then we find an equation for the derivative of f. Then we solve to find the value of x for which the derivative is 0.

```
W File Edit Insert Action
[icons]
□
                              [icon]
Define f(x)=x³/9 + x²/6 −2·x−1
                         done
d/dx(f(x))
                    x²+x−6
                    ──────
                      3
solve(ans=0,x)
                  {x=−3,x=2}
□

Alg   Standard Cplx Rad [battery]  f
```

In this case, we get the answers x=-3, and x=2. We can find the value of the original function (the y-coordinate of the maximum or minimum point), by substituting x in the original equation:

```
W File Edit Insert Action
[icons]
define f(x)=x³/9 + x²/6 −2·x−1
d/dx f(x)
                    x²+x−6
                    ──────
                      3
solve(Ans=0,x)
                  {x=−3,x=2}
f(x)|x=−3
                       7
                       ─
                       2
f(x)|x=2
                      31
                    − ──
                       9
□

Alg   Standard Cplx Rad [battery]
```

We should note in passing that this technique only finds the local maximum or minimum. One glance at the full curve suggests that the global maximum and minimum are infinite:

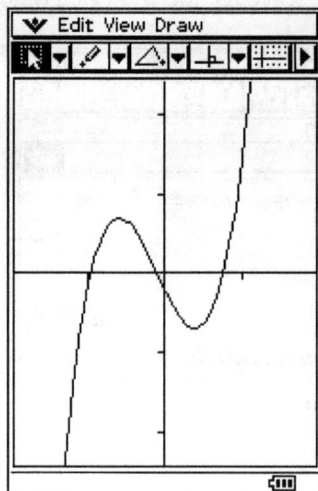

Now our technique detected the location of a local maximum and a local minimum, but it was not able to tell which was which (of course, one look at the graph told us). But there is a way to distinguish the two situations using calculus rather than the eyeball.

Look at the tangent to the derivative curve at the points where it crosses the x-axis:

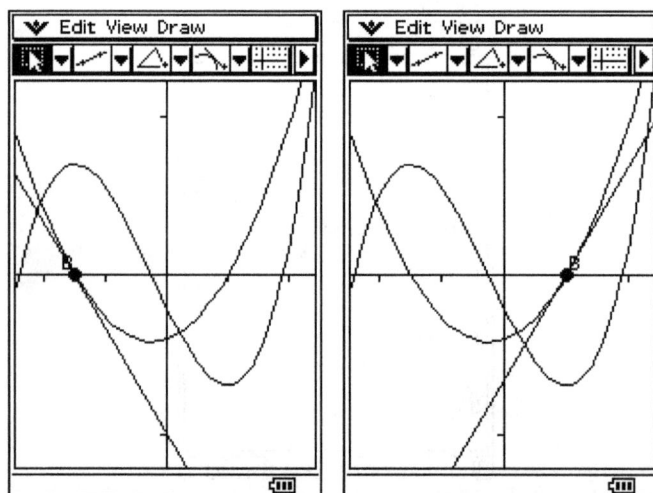

We see that in the location which corresponds to a local maximum, the slope of this tangent is negative, whereas in the location which corresponds to a minimum of the original curve, this tangent has a positive slope.

We can determine whether a particular solution is a maximum or a minimum by taking a second derivative (the slope of the tangent of the derivative curve). Substituting x=–3 gives a negative value, indicating this is a maximum. Substituting x=2 gives a positive value, indicating this is a minimum:

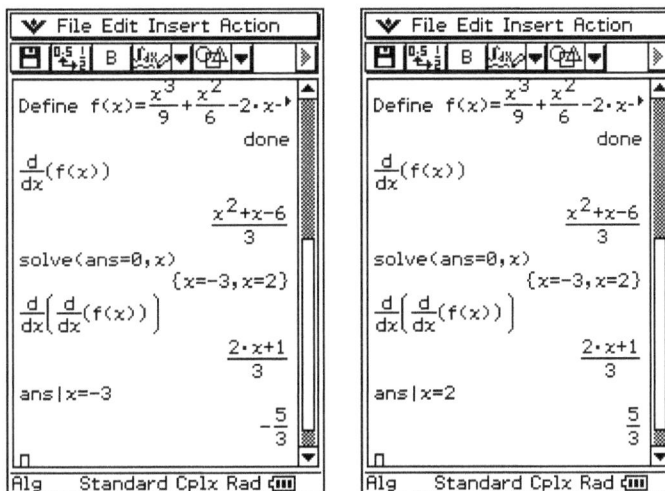

EXERCISES

1. Draw the function $\dfrac{x^3}{3} - x^2 + x - 1$ in a geometry window. Does it have any local maxima or minima?

2. Differentiate the above function and solve for $\dfrac{df}{dx} = 0$. Plot the derivative. What do you notice about the derivative curve with respect to the x-axis?

3. Evaluate $\dfrac{d}{dx}\left(\dfrac{df(x)}{dx}\right)$ at the point where $\dfrac{df}{dx} = 0$. This point is called an *inflection point* of the original curve.

4. Find a local minimum of the curve $x^2 + 2x - 3$

5. Find all local maxima and minima of the curve $x^3 - 2x^2 - 4x + 3$

# The Best Ferry Angle

A kayaker paddling across a river will point his boat somewhat upstream so that he can counteract the effect of the current trying to push him downstream. The angle he points the boat depends on the strength of the current. This maneuver is called a "ferry".

We can illustrate the situation with vectors. Let's assume for the sake of argument that the kayaker paddles at 1 m/s, and that the river current is 0.75 m/s:

First draw a vector representing the kayak's velocity. Set it's length to 1. Now create a second vector representing the current. Set its length to be 0.75 and its direction to be vertical.

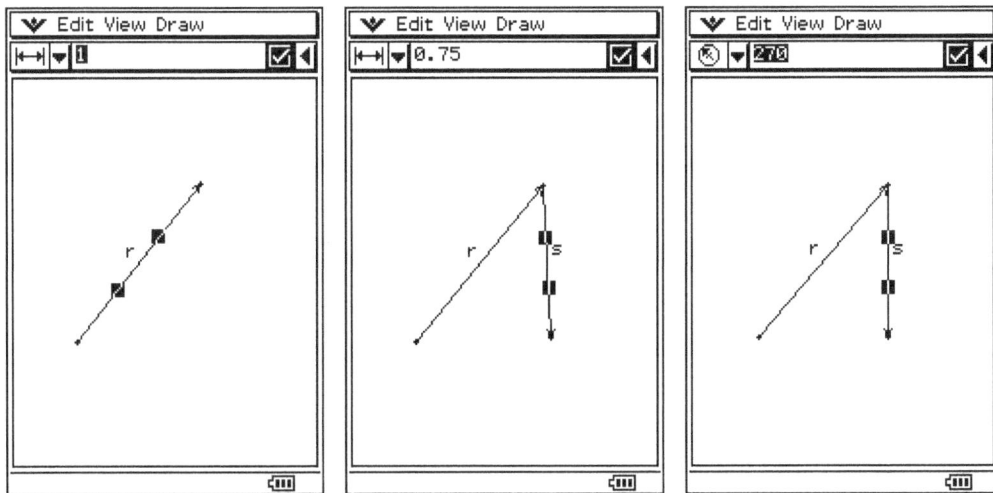

Now draw a vector representing the resultant motion (joining the start of r to the end of s.) If the kayaker has pointed his boat correctly, this vector will be horizontal. Set it to horizontal, then you can read off the correct angle:

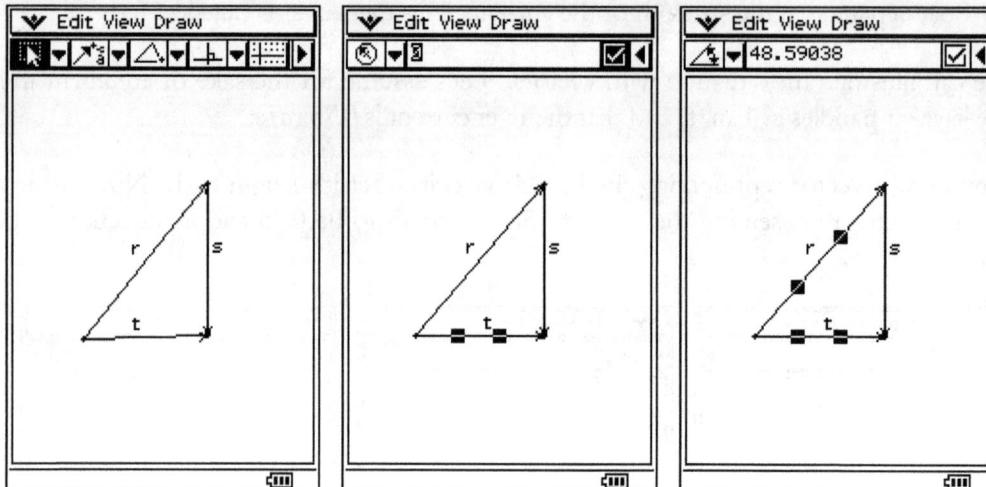

What happens to this angle when the current speed is 0.9m/s?

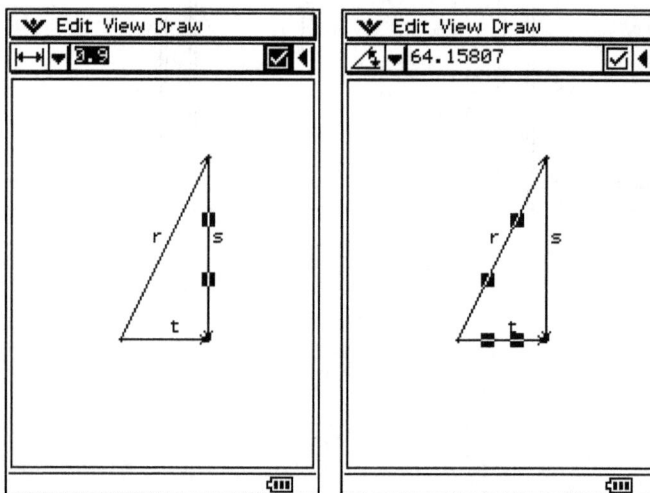

Clearly as the speed of the current gets closer to the speed of the kayaker, he needs to point his boat more and more upstream to match it. This of course means that he will take longer and longer to cross the river.

Let's assume we have a stream 1 meter wide (a narrow stream, but one which makes the math simpler), how long will it take him to cross if the current speed is x m/s, where x<1?

Simple trigonometry shows us that the cross stream velocity is cos(θ)  , where $\theta = \sin^{-1} x$

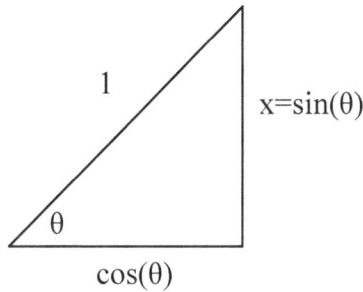

Hence the time to get across the current is $\dfrac{1}{\cos(\theta)}$

We can look at a graph of that:

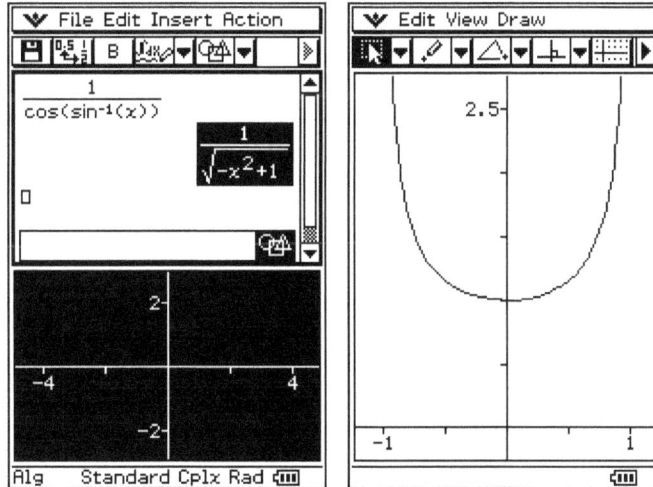

Of course, as x approaches 1, the poor kayaker is pointing more or less straight upstream and is making no progress at all. It will take him a long time to cross the river.

What about the situation where the current is faster than the kayaker is capable of paddling? Clearly it is no longer possible for him to ferry across the river without drifting downstream. In this situation, what is the best angle for him to point?

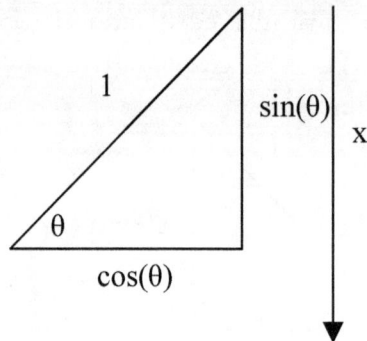

We assume an angle $\theta$ such that the cross stream component of the motion is y. In that case the upstream component is

$$\sin(\theta)$$

The ground speed (downstream drift speed) of the kayak is

$$x\text{-}\sin(\theta)$$

Now the total time spent in crossing the river, is the width of the river divided by the cross-stream speed:

$$\frac{1}{\cos(\theta)}$$

Hence the total drift is:

$$\left(x - \sin(\theta)\right) \cdot \frac{1}{\cos(\theta)}$$

So what is the best value for $\theta$?

The best value, one might argue, is the one which allows the kayak to drift downstream the least. If you are just upstream from a waterfall, you might argue this quite strenuously. So what is the magic angle?

First, we define f(θ) to be the downstream drift, then we differentiate f (a little simplification helps):

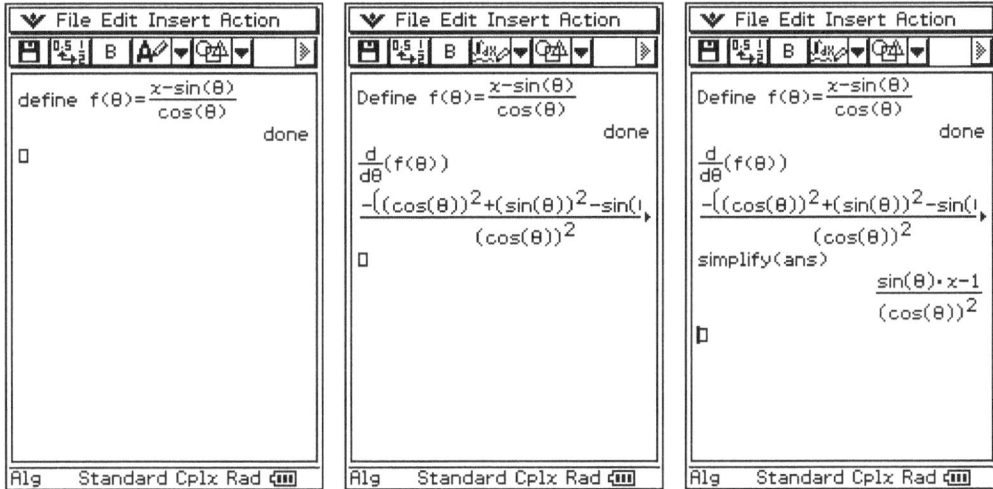

```
W File Edit Insert Action
[icons]                    »

define f(θ)=(x−sin(θ))/cos(θ)
                        done
□




Alg    Standard Cplx Rad (iii)
```

```
W File Edit Insert Action
[icons]                    »

Define f(θ)=(x−sin(θ))/cos(θ)
                        done
d/dθ(f(θ))
−((cos(θ))²+(sin(θ))²−sin(!
        (cos(θ))²        ▸
□



Alg    Standard Cplx Rad (iii)
```

```
W File Edit Insert Action
[icons]                    »

Define f(θ)=(x−sin(θ))/cos(θ)
                        done
d/dθ(f(θ))
−((cos(θ))²+(sin(θ))²−sin(!
        (cos(θ))²        ▸
simplify(ans)
            (sin(θ)·x−1)/(cos(θ))²

□

Alg    Standard Cplx Rad (iii)
```

Now we need to solve $\dfrac{df}{d\theta} = 0$ for θ:

This has a couple of solutions, the relevant one being the second (the first is all angles bigger than π). You can drag and drop the solution into a Geometry Window.

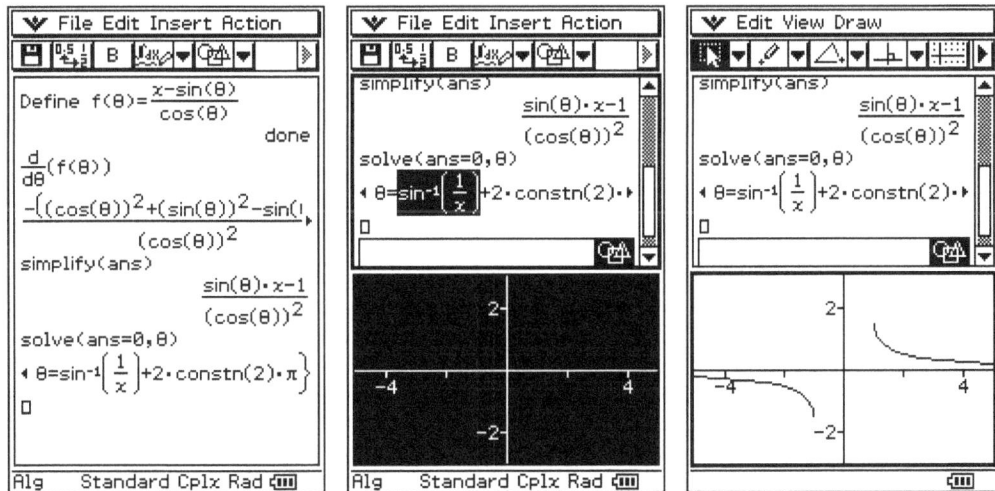

```
W File Edit Insert Action
[icons]                    »

Define f(θ)=(x−sin(θ))/cos(θ)
                        done
d/dθ(f(θ))
−((cos(θ))²+(sin(θ))²−sin(!
        (cos(θ))²        ▸
simplify(ans)
            (sin(θ)·x−1)/(cos(θ))²

solve(ans=0,θ)
◄ θ=sin⁻¹(1/x)+2·constn(2)·π}
□

Alg    Standard Cplx Rad (iii)
```

```
W File Edit Insert Action
[icons]                    »

simplify(ans)
            (sin(θ)·x−1)/(cos(θ))²
solve(ans=0,θ)
◄ θ=sin⁻¹(1/x)+2·constn(2)·▸
□
                        [icon]

          2-


  −4               4

         −2-

Alg    Standard Cplx Rad (iii)
```

```
W Edit View Draw
[icons]                    ►

simplify(ans)
            (sin(θ)·x−1)/(cos(θ))²
solve(ans=0,θ)
◄ θ=sin⁻¹(1/x)+2·constn(2)·▸
□
                        [icon]

          2-


  −4               4

         −2-

                        (iii)
```

Putting a point on the curve lets us extract a value.  For example at x = 1.5 (river current 1.5 times kayak speed, we see the optimal angle is 0.729728 radians.  Or 41.81 degrees.

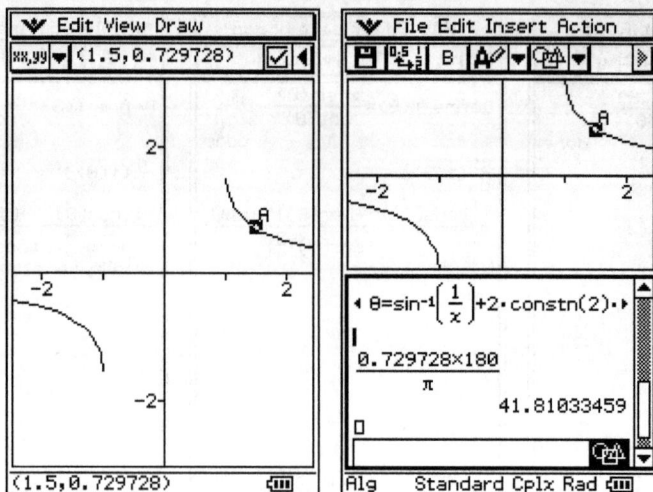

Collecting our two results, we have the following solution to the ferry angle problem:

Given boat speed 1, and river speed x, the best ferry angle is

$$\theta = \sin^{-1} x \quad \text{where x<1}$$

$$\theta = \sin^{-1} \frac{1}{x} \quad \text{where x>1}$$

The interesting symmetry of these two results may be interpreted in the following rule of thumb:

You use the same ferry angle if the speed of the current is twice your speed as you do if your speed is twice that of the current.

1. A lidless cardboard box is made by cutting squares out of the corners of a rectangular piece of cardboard measuring 100cm by 75cm, then folding the sides up. What is the maximum volume of box which can be made in this way?

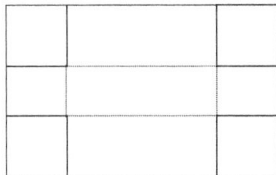

2. A farmer wants to create a rectangular enclosure using chicken wire. One side of the enclosure is formed by the wall of a barn. He has 100' of wire to form the other 3 sides. What dimensions would give him the greatest area enclosed. What would the area be?

# Rate Equations

## A Growth Equation

In a situation of unconstrained growth, a population (of men, of mice, of microbes) may have a growth rate proportional to the size of the population. For example, let's assume a single lemming pair has on average 12 offspring per year, and the offspring are immediately fertile. Also assume that the breeding goes on all year (not too far from the truth for lemmings). Then the birthrate is 0.5 lemmings per head of population per month.

Then if P(t) is the population of lemmings at time t, then the growth rate of the population is 0.5*P. That is, the rate of growth of the population is proportional to the size of the population itself – the more lemmings you have, the more new lemmings you get each month.

If we were looking at a graph of the population, the rate of change of population at time t would be the slope of the tangent to the graph at t. So, we can write the following about P(t):

$$\frac{dP(t)}{dt} = 0.5 * P(t)$$

So P(t) is a function whose derivative is half of the original function. Can we guess what P(t) is?

Well, we know one function whose derivative is itself:

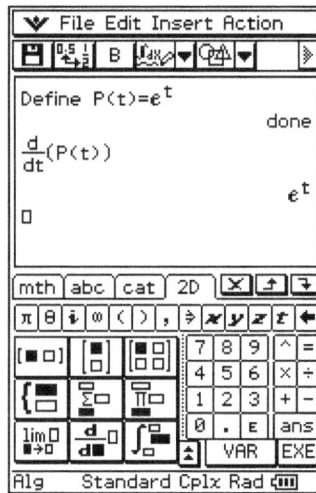

```
▼ File Edit Insert Action
[icons toolbar]
Define P(t)=e^t
                          done
d
──(P(t))
dt
                              e^t
□
mth abc cat 2D  [toolbar]
π θ i ∞ ( ) , ⇒ x y z t ←
[■□] [■] [■□] 7 8 9 ^ =
                4 5 6 × ÷
{■  Σ□  Π□  1 2 3 + −
            0 . E ans
lim□ d□ ∫□  VAR EXE
Alg    Standard Cplx Rad ▣
```

Can we change the function a little to make the derivative half of the original function?

How about multiplying the original function by a half?

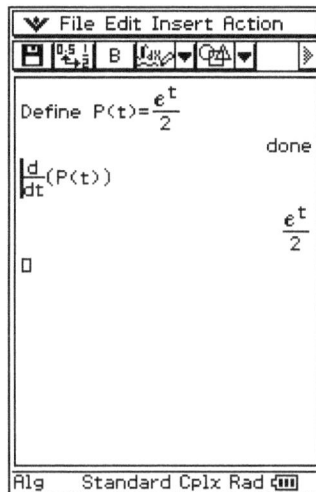

```
▼ File Edit Insert Action
[icons toolbar]
Define P(t)=e^t/2
                          done
d
──(P(t))
dt
                              e^t
                              ──
                               2
□




Alg    Standard Cplx Rad ▣
```

No, this doesn't work. We've halved both P and its derivative so we still have the derivative of P equal to P.

How about halving t in the original function?

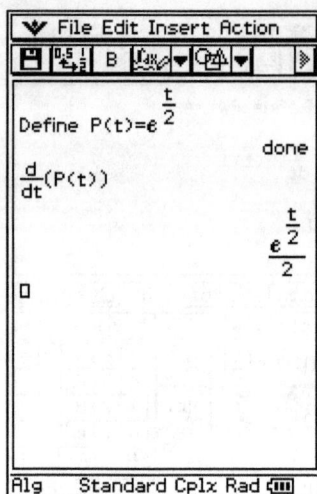

```
▼  File Edit Insert Action
[icons]                        ▷

                          t
                          ─
Define P(t)=e  2
                              done
 d
─(P(t))
dt
                             t
                             ─
                           e 2
                           ───
                            2
□



Alg    Standard Cplx Rad ▥
```

We see that indeed $\dfrac{dP(t)}{dt} = 0.5 * P(t)$. Is this the only such function? What other functions can you find that also have this property?

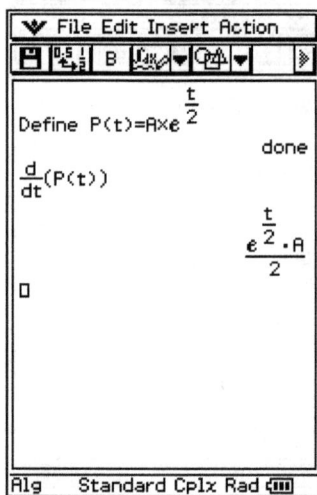

```
▼  File Edit Insert Action
[icons]                        ▷

                          t
                          ─
Define P(t)=A×e 2
                              done
 d
─(P(t))
dt
                            t
                            ─
                          e 2·A
                          ─────
                            2
□



Alg    Standard Cplx Rad ▥
```

We have a whole family of possible functions for our lemming population. What is the population at time t=0?

So if we start with 2 lemmings, then A should be 2. In order to drag and drop a graph into geometry, the independent variable needs to be x:

One look at the population graph convinces you of the need for snowy owls and arctic foxes.

EXERCISES

1.  The rate of radioactive decay is proportional to the quantity of radioactive material present. This can be written: $\dfrac{dR}{dx} = -\lambda R$, where R is the quantity of radioactive material, x is time, and $\lambda$ is the radioactive constant. Find a family of functions R(x) which satisfy the above equation.

2.  The half life of a radioactive substance is the value for x such that $R(x) = \dfrac{R(0)}{2}$. If a substance has a half life of 25 years, what is the equation for R(x)?

## Speeds and Slopes

For the first 5 seconds of its motion, a bicycle moves so that its displacement (distance from starting point) in meters as a function of time measured in seconds f(x) follows this formula:

$$f(x) = \frac{x^3}{27} - \frac{x^2}{12} + \frac{x}{4}$$

What speed is it going after 3 seconds?

Let's look at a graph of the function:

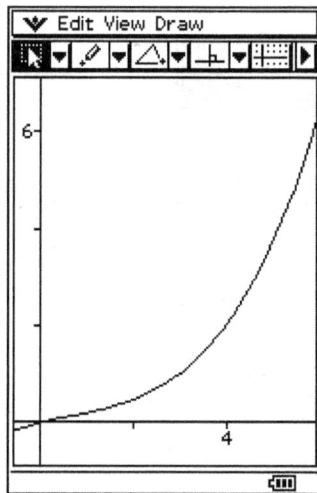

Speed is distance traveled / time taken. We could look at the point on the curve with x coordinate 3 (turning on grid points aids in accurate point placement).

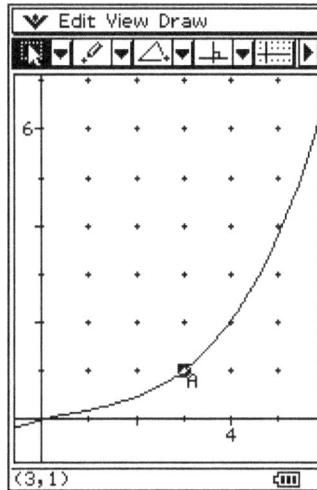

After 3 seconds, the bicycle has gone 1 meter, so its average speed over that time is 0.3333 m/s. That is the slope of the chord joining the origin to A:

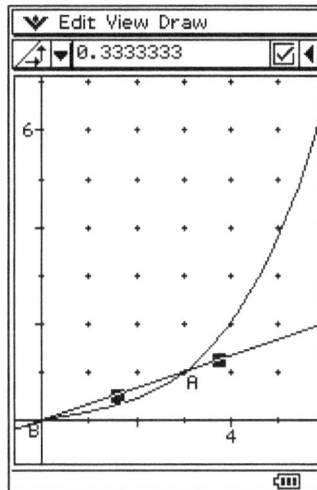

This is the average speed, but it is clear from the graph that the bicycle is accelerating (getting faster) all the time. We can see this by examining the slope of the chords from x=1 to x=3 and from x=2 to x=3:

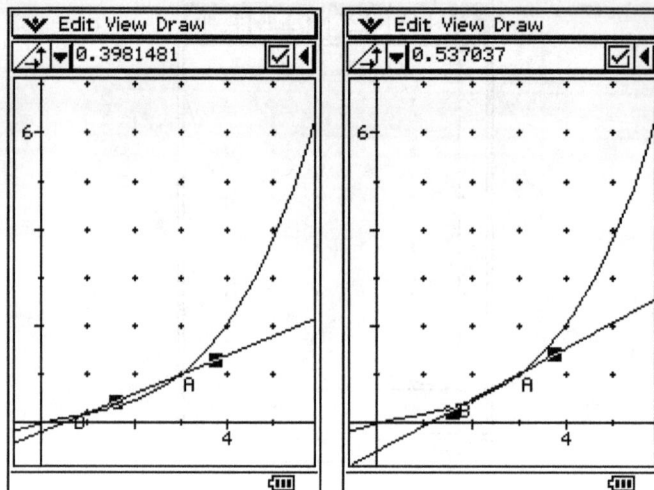

Clearly, the closer we make B come to A, the closer we get to an accurate representation of the speed of the bicycle after 3 seconds. And if we use the tangent instead of the chord, then the slope will be exactly the speed we want:

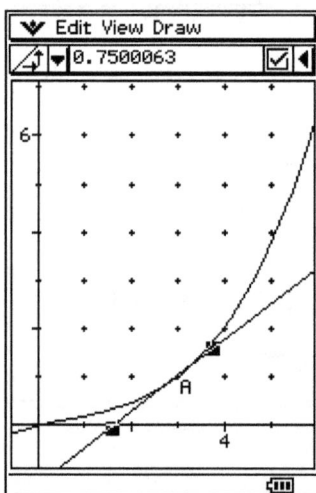

## Newton's 3rd Law

Newton's 3$^{rd}$ law says force = mass times acceleration:

F=ma

Now, to a ball of mass m thrown upwards on the earth, the force is a constant: mg. So a=g.

And it turns out that g≈9.8 (when measured in SI units).

The acceleration of an object is just the rate of change of its velocity. That is, if we looked at a graph of velocity against time, the slope of the tangent at time t would be the acceleration at time t.

If v(t) is the velocity function, we therefore know that:

$$\frac{dv(t)}{dt} = g$$

What function do we know, whose derivative is a constant?

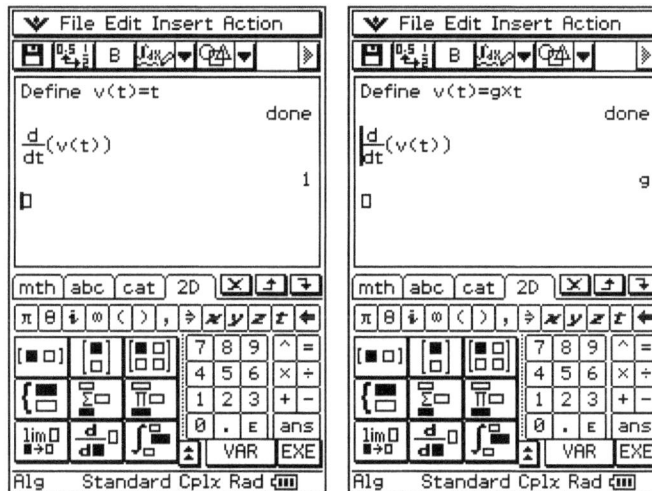

```
W File Edit Insert Action          W File Edit Insert Action
[icons]                      >>    [icons]                      >>
Define v(t)=t                      Define v(t)=gxt
                      done                               done
d                                  d
──(v(t))                           ──(v(t))
dt                                 dt
                         1                                  g
□                                  □

mth abc cat  2D  X ± ↓             mth abc cat  2D  X ± ↓
π θ i ∞ ( ) , ⇒ x y z t ←          π θ i ∞ ( ) , ⇒ x y z t ←
[■□] [■/□] [■□/□□] 7 8 9 ^ =       [■□] [■/□] [■□/□□] 7 8 9 ^ =
                 4 5 6 × ÷                          4 5 6 × ÷
{■□} Σ□ π□       1 2 3 + -         {■□} Σ□ π□       1 2 3 + -
lim□ d□ ∫□□      0 . E ans         lim□ d□ ∫□□      0 . E ans
■→□ d■          ± VAR EXE          ■→□ d■          ± VAR EXE
Alg   Standard Cplx Rad ▦          Alg   Standard Cplx Rad ▦
```

The derivative of g.v(t) is equal to g. Are there any other functions which have g as their derivative?

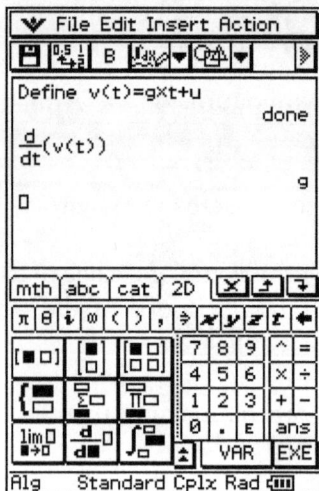

```
▼ File Edit Insert Action
[ ][  ][B][   ][   ]          [≫]
Define v(t)=g×t+u
                        done
d
──(v(t))
dt
                           g
□

mth  abc  cat   2D  [X][↴][↯]
π 8 i ∞ ( ) , ⇒ x y z t ←
[■□] [■] [■□]  7 8 9 ^ =
     [□] [□□]  4 5 6 × ÷
{■   Σ□   π□   1 2 3 + −
 □           0 . E ans
lim□  d□  ∫□  [↕] VAR EXE
■→□  d■  □
Alg    Standard Cplx Rad ▭
```

Yes, as the derivative of a constant is 0, you can add on any constant.

Evaluate the velocity function at t=0:

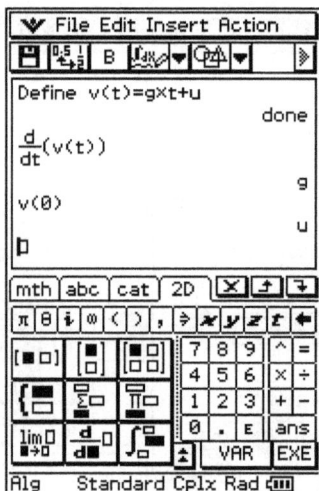

```
▼ File Edit Insert Action
[ ][  ][B][   ][   ]          [≫]
Define v(t)=g×t+u
                        done
d
──(v(t))
dt
                           g
v(0)
                           u
□

mth  abc  cat   2D  [X][↴][↯]
π 8 i ∞ ( ) , ⇒ x y z t ←
[■□] [■] [■□]  7 8 9 ^ =
     [□] [□□]  4 5 6 × ÷
{■   Σ□   π□   1 2 3 + −
 □           0 . E ans
lim□  d□  ∫□  [↕] VAR EXE
■→□  d■  □
Alg    Standard Cplx Rad ▭
```

We see that the velocity at time t is:

$$v(t)=g.t+v(0)$$

Now, velocity itself is the rate of change of position. If s(t) is the height of the ball at time t, then

$$\frac{ds(t)}{dt} = v(t) = g\,t + u$$

We are looking for a function whose derivative is the linear function g.t+u. Try to find a function s(t) whose derivative is g.t+u.

Can you find any other functions with this derivative?

Now what is the displacement at time t=0:

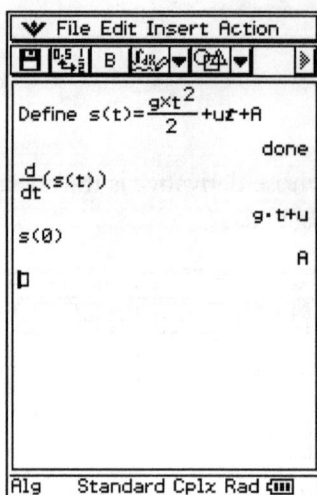

```
▼ File Edit Insert Action
🖫 🖫 B 🖫▼ 🖫▼          ≫

Define s(t)=gxt²/2 +ut+A
                        done
d/dt(s(t))
                      g·t+u
s(0)
                         A
▯

Alg    Standard Cplx Rad ▥
```

We can see that the height of the ball at time t is:

$$s(t) = \frac{g}{2}t^2 + v(0) \cdot t + s(0)$$

We can graph the height of a ball against time, if we set values for the gravitational constant (approximately 9.8 m/s/s) and for initial velocity and initial height. Let's use 8 for initial velocity and 0 for initial height:

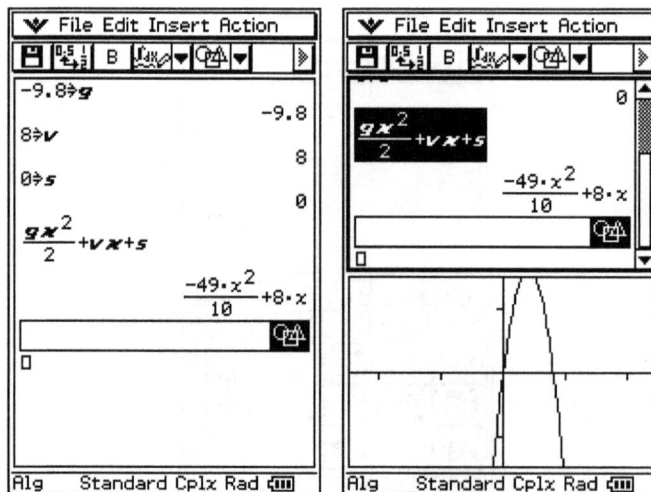

```
▼ File Edit Insert Action
🖫 🖫 B 🖫▼ 🖫▼          ≫
-9.8⇒g
                      -9.8
8⇒v
                         8
0⇒s
                         0
gx²/2 +vx+s
              -49·x²/10 +8·x
                       🖫
▯

Alg    Standard Cplx Rad ▥
```

```
▼ File Edit Insert Action
🖫 🖫 B 🖫▼ 🖫▼          ≫
                         0 ▲
gx²/2 +vx+s
              -49·x²/10 +8·x
                       🖫
▯                        ▼


Alg    Standard Cplx Rad ▥
```

Let's say we were interested in computing the length of time before the ball hit the ground. To do this we need to find the x-value where the graph crosses the axis. Zooming in on the graph can give us an approximate value:

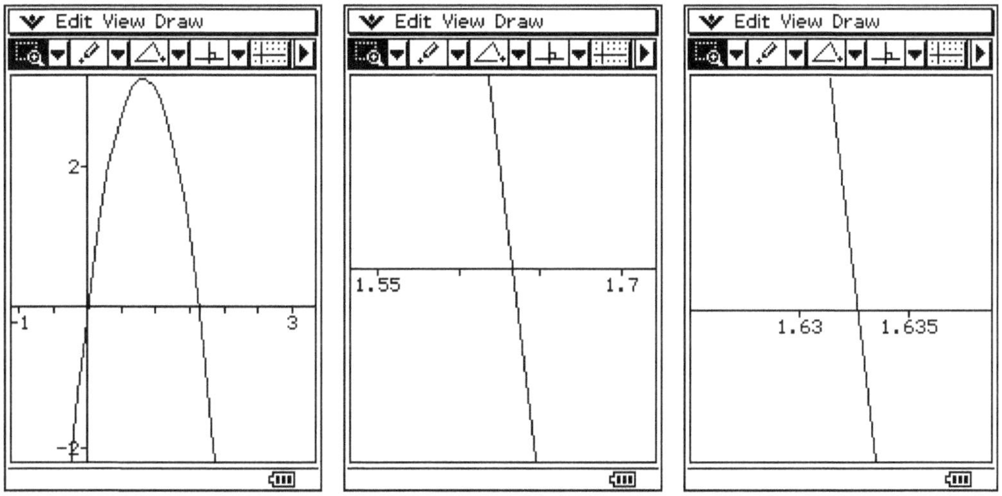

Another approach is to solve for height=0 in the eActivity window:

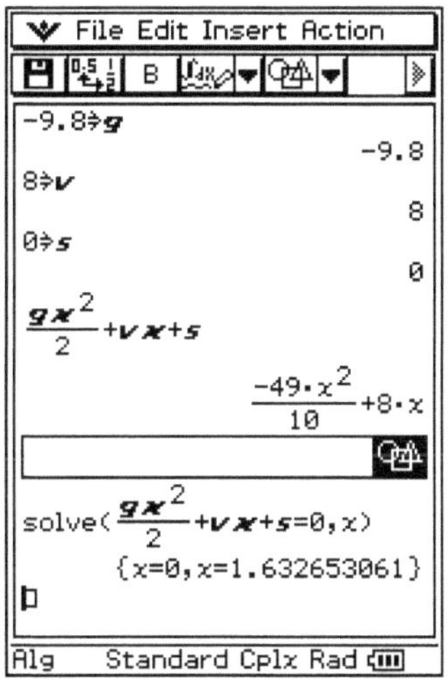

We see there are 2 solutions, the first solution at time 0, just before the ball is thrown, and the second solution at time 1.6327 seconds.

EXERCISES

1.  A climber falls from 100m. How long does it take him to reach the ground?

2.  For a shell fired from a gun at angle $\theta$ to the horizontal with muzzle velocity v, the vertical component of its velocity is v.sin($\theta$)

    a.  Derive a formula for the length of time before the shell hits the ground.

    b.  The horizontal component of the shell's velocity is v.cos($\theta$). What horizontal distance will the shell have traveled before it hits the ground?

    c.  What angle $\theta$ gives a maximum range?

3.  Springs satisfy Hooke's law, which says that if f(x) measures displacement from the natural length of the spring, then $\dfrac{d}{dx}\left(\dfrac{d}{dx}(f(x))\right) = -kf(x)$, where k is a constant measuring the stiffness of the spring.

    a.  Find a function which satisfies: $\dfrac{d}{dx}\left(\dfrac{d}{dx}(f(x))\right) = -f(x)$

    b.  Find a function which satisfies: $\dfrac{d}{dx}\left(\dfrac{d}{dx}(f(x))\right) = -4f(x)$

# Approximating Functions

Polynomials are nice simple functions. It can be useful to find a polynomial which approximates a given non-polynomial function. The approximation is always local, not global. That is we specify the x value which we are interested in and look for a polynomial approximation which is close to this x-value. Typically the approximation will be less good the further away from the x value we get.

The function taylor() in eActivity produces a polynomial of given degree which approximates a function at a point. The syntax is as follows:

taylor(**function**, **variable**, **degree**, **x_location**)

This creates a polynomial approximation to the **function** local to **variable**=**x_location** of the given **degree**.

Let's study the function:

$$f(x) = \frac{5}{\sqrt{1 + (x+1)^2}}$$

We'll create a first degree approximation at x=0:

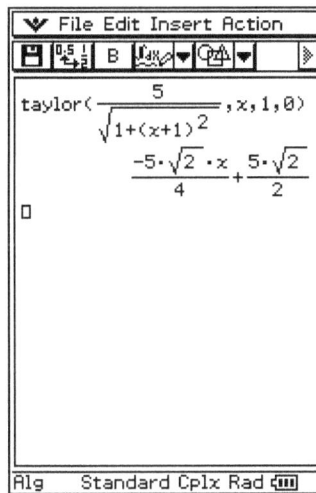

We see this is the equation of a line. We can drag and drop the original function into a Geometry Strip, then drag in the equation of the linear approximation:

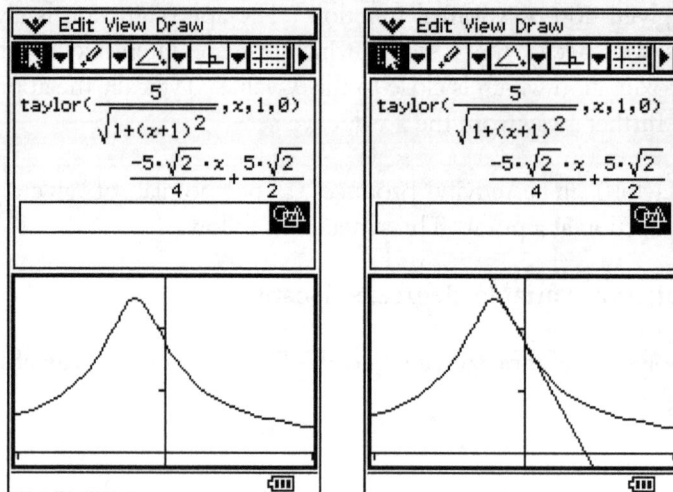

We see the linear approximation is the tangent at x=0.

Let's try a second order approximation:

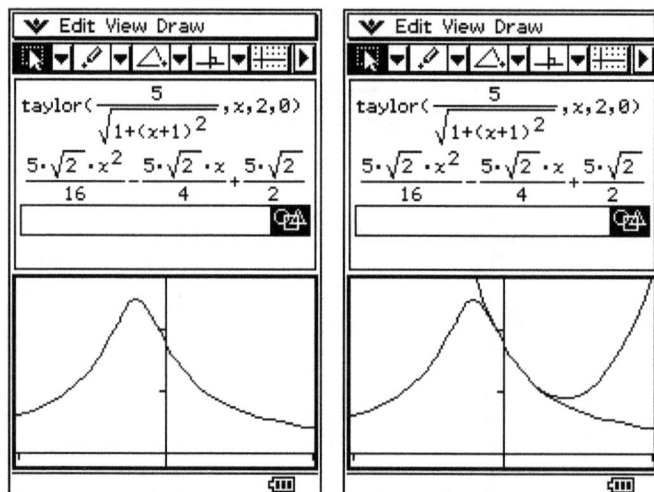

We get a quadratic function which matches the original curve at x=0.

Higher order approximations give higher order polynomials which match the original curve closer near the y axis:

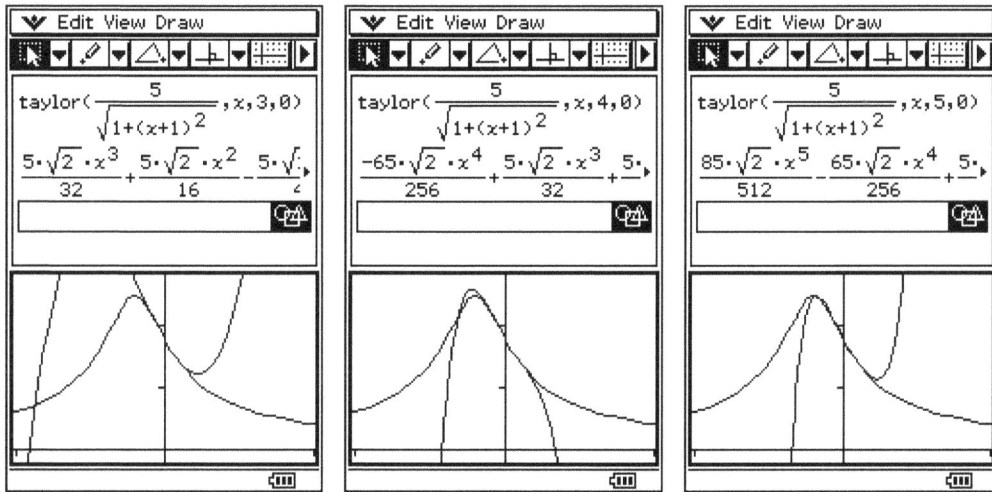

Let's take a closer look at these approximations. The first order polynomial is just the tangent at x=0. It should therefore come as no surprise that the coefficient of x is the slope of the tangent of the original curve at x=0, and the constant term is the value of the original function at x=0.

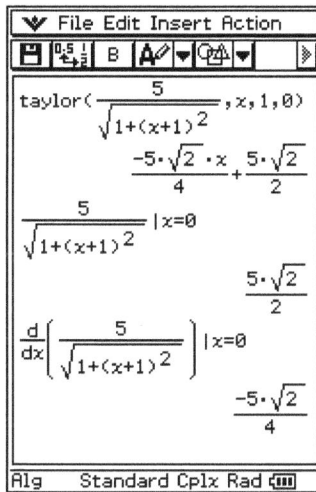

Let's define f(x) to be our original function and p(x) to be our polynomial approximation of degree 2:

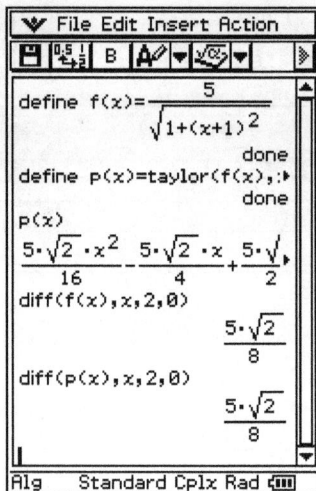

We see that the $2^{nd}$ derivative of p(x) coincides with the $2^{nd}$ derivative of f(x).

Let's do the same with the next approximation:

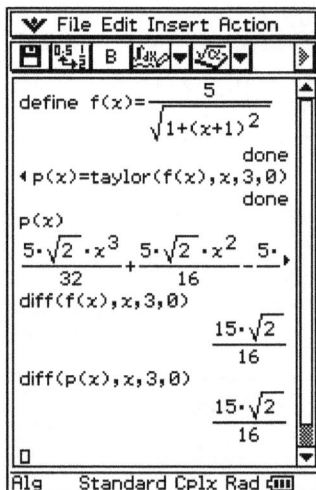

Can you work out a formula for constructing the nth coefficient of the Taylor Polynomial?

Here is a hint:

```
 ▼ File Edit Insert Action
 ▣ ▤ B ▨▼ ▨▼      ⟫
─────────────────────────
diff(x^n,x,2)
                  n·(n-1)·x^(n-2)
diff(x^n,x,3)
              n·(n-2)·(n-1)·x^(n-3)
diff(x^n,x,4)
         n·(n-3)·(n-2)·(n-1)·x^(n-4)
 ▫
─────────────────────────
Alg   Standard Cplx Rad ▥
```

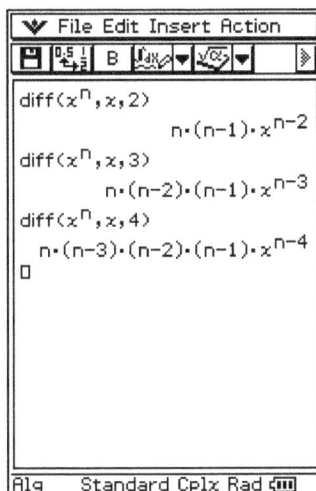

Hence, we can see that:

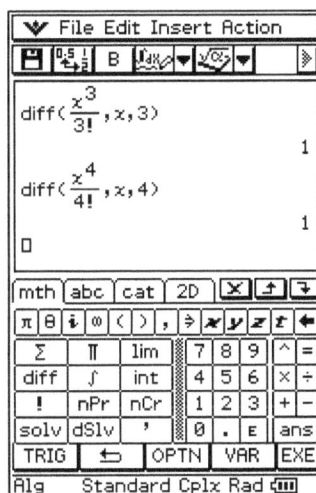

```
 ▼ File Edit Insert Action
 ▣ ▤ B ▨▼ ▨▼      ⟫
─────────────────────────
diff( x^3/3! ,x,3)
                              1
diff( x^4/4! ,x,4)
                              1
 ▫
─────────────────────────
 mth  abc  cat  2D  [X][↑][↴]
 π θ i ∞ ⟨ ⟩ , ÷ x y z t ←
  Σ    Π   lim ║ 7 8 9 ║ ^ =
 diff   ∫   int ║ 4 5 6 ║ × ÷
  !   nPr  nCr ║ 1 2 3 ║ + −
 solv dSlv  '  ║ 0 . E ║ ans
 TRIG   ⇆   OPTN  VAR  EXE
─────────────────────────
Alg   Standard Cplx Rad ▥
```

**127**

So if we desire the third derivative of our polynomial to be A, for example, we can set the coefficient of $x^3$ to be $\dfrac{A}{3!}$:

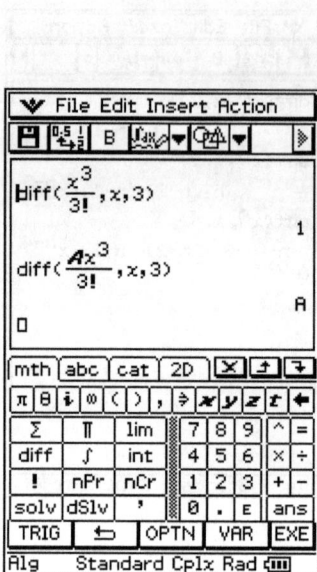

```
▼ File Edit Insert Action
[icons toolbar]

diff( x³/3! ,x,3)
                                    1

diff( Ax³/3! ,x,3)
                                    A
□

mth  abc  cat  2D  [X][±][∓]
π  θ  i  ∞  (  )  ,  ⇒  x  y  z  t  ←
Σ    Π    lim    7  8  9  ^  =
diff  ∫    int    4  5  6  ×  ÷
 !   nPr  nCr    1  2  3  +  −
solv dSlv  '     0  .  E  ans
TRIG    ⇆    OPTN    VAR    EXE
Alg    Standard Cplx Rad
```

EXERCISES

1. Create the terms up to $5^{th}$ order in the Taylor series for sin(x) at x=1. Plot them against the graph of sin(x).

2. Create the terms up to $4^{th}$ order in the Taylor series for cos(x) at x=1. Plot them against the graph of cos(x).

3. Create a polynomial whose value at x=0 is 2, whose first derivative at x=0 is –1 and whose second derivative at x=0 is 0.5.

# Integral Calculus

Integral Calculus involves finding the area under a curve.

Why is this such a big deal?

Again, two answers:

1. Finding areas is pretty important in real life as are related tasks such as finding weight, volume, energy, etc.

2. Integration turns out to be the inverse operation to differentiation.

## Area Under a Line

Let's start with a very simple sort of a curve – we'll look at the area under a straight line. We first draw the line AB, then draw segments AC and CD, setting the direction of CD to be vertical.

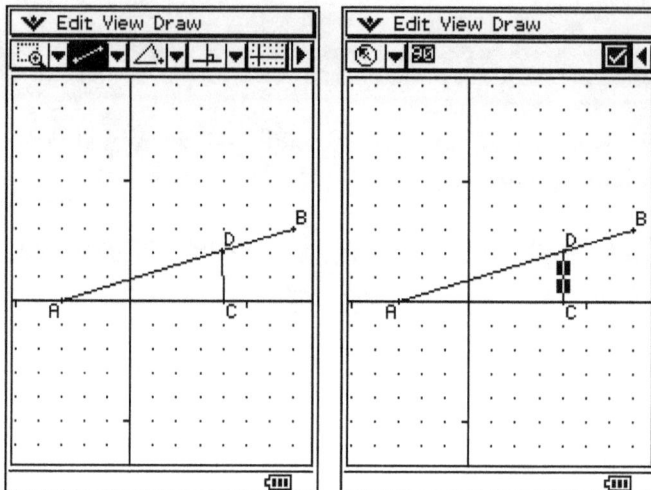

We will look at the area of triangle ACD as D runs along AB. To do this we select the point D and the line AB, then select Animate / Add Animation from the Edit menu:

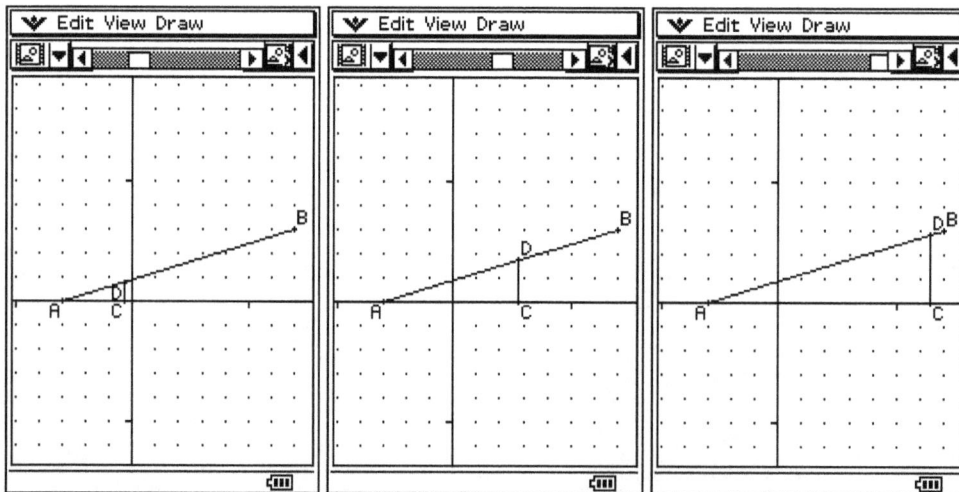

We can display the area of the triangle in the measurement box with points A,C and D selected. Pressing the Table button creates a list of these values through the range of the animation.

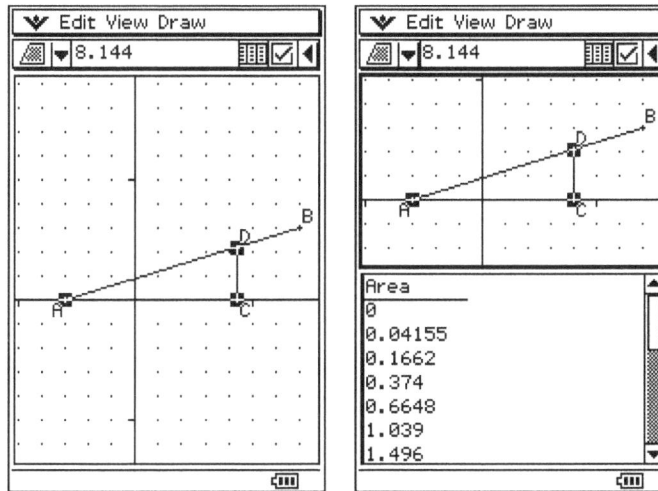

We'd like to graph area against the x-coordinate of C. To do this, we need to tabulate the coordinates of C, then move the x column to the front.

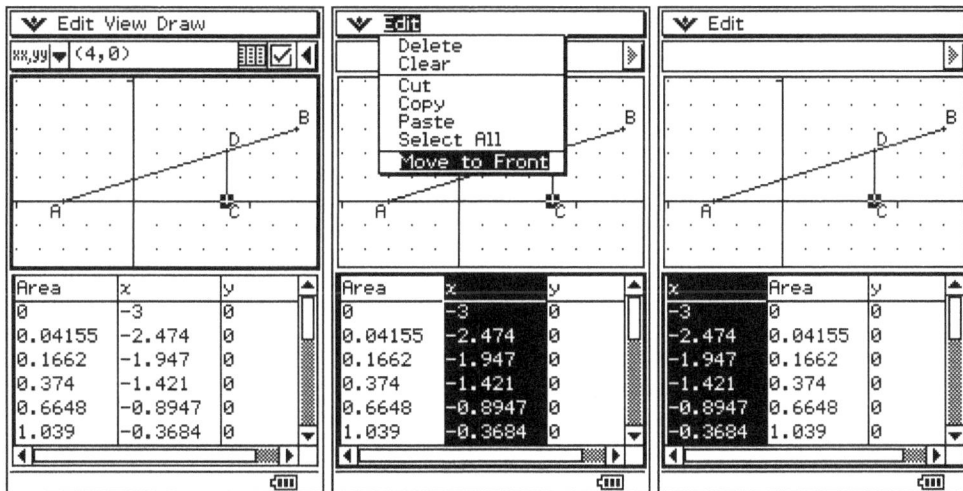

We can now select both the x column and the Area column and drag them into the geometry window. A plot of the values is created:

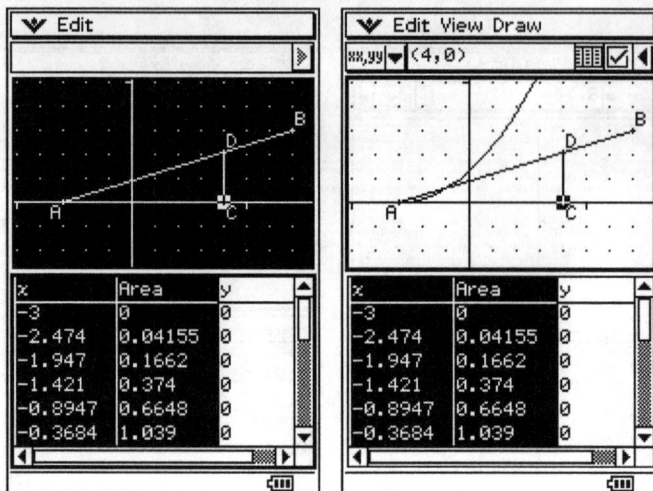

## Name That Curve

So what curve do we have? It does look like a quadratic, and since we are dealing with areas which tend to be squares, it would seem that a quadratic might be a good place to start. Let's draw the function x^2 and see how it looks:

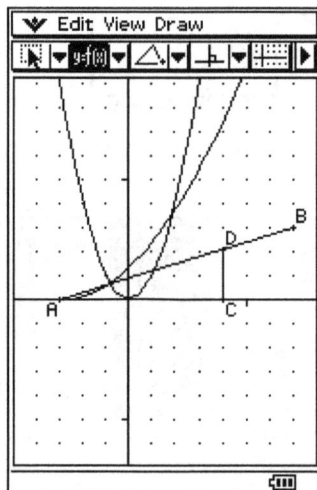

Well, it looks too steep and its minimum is in the wrong place. We can sort the minimum easily, by dragging:

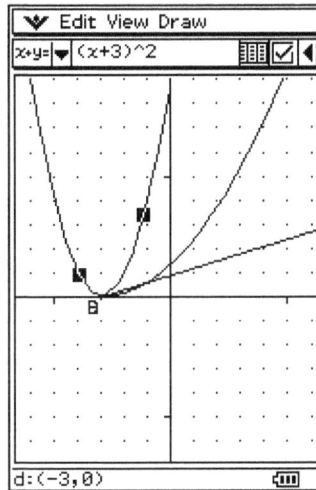

We should be able to sort the steepness out by multiplying. Let's do a little trial and error:

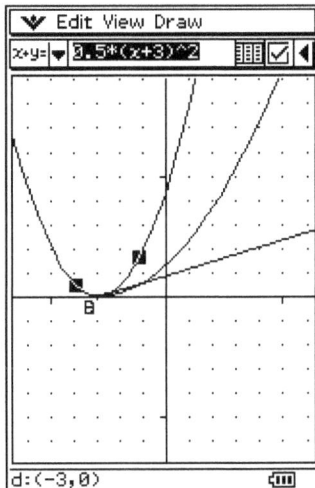

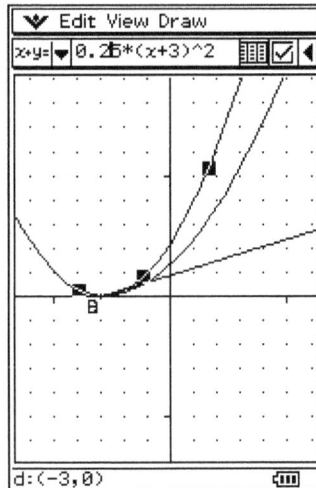

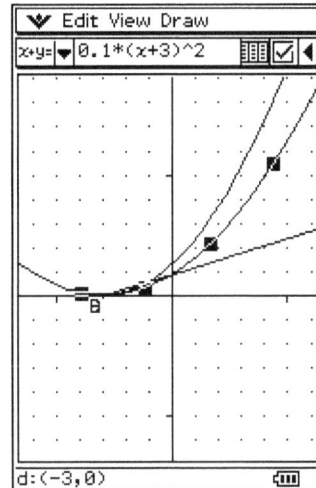

We are somewhere between 0.1 and 0.25:

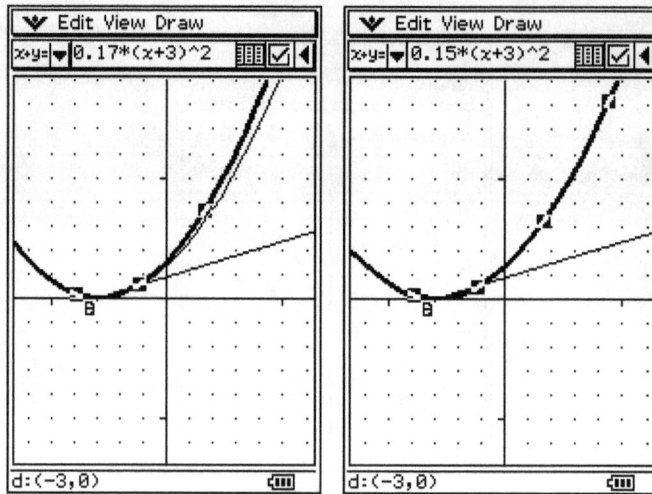

And indeed, it looks like the answer is 0.15.

EXERCISES

1. For the example in the text, write down an equation for $|DC|$ in terms of the x coordinate of C.

2. Write down a formula for the area of triangle ADC in terms of x. Compare with the expression derived in the text.

## Finding the Area Under a Curve

Let's imagine we want to calculate the area under the curve:

$$y = 0.05x^3 - 0.05x^2 - 0.5x + 3$$

between the points x=-4 and x=4.

We can attempt this by successive approximation, looking at collections of polygons which approximate the desired area.

First, let's draw the curve - to facilitate placement of our polygons, we have the grid and axes turned on:

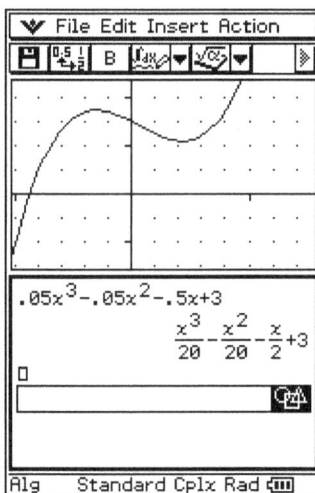

Now we create our first approximation: the quadrilateral ABCD

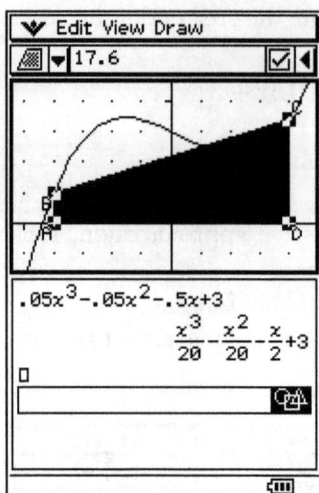

Care is needed when drawing the small side (AB above) to make sure that point A does not snap to the curve, but that point B does!

With the four sides of the quadrilateral selected (or the four corners), the area may be read off from the Measure Box. (It can also be shaded, using **Edit / Shade On/Off**)

Before proceeding, let's stash our area in eActivity. We'll make approximations to the area with successively thinner trapezoids. We'll save the width of the trapezoid along with its area. For later convenience, we'll stash these in a matrix:

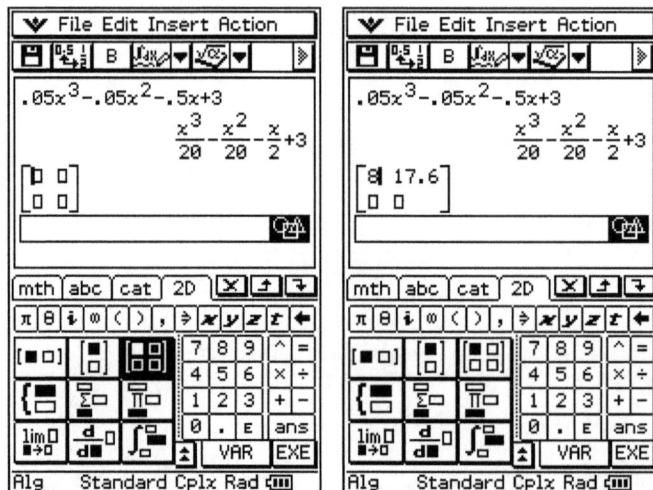

Now for our next approximation, we need to delete the line BC and replace it by a pair of lines BE EC, where E is a new point on the curve at x=0:

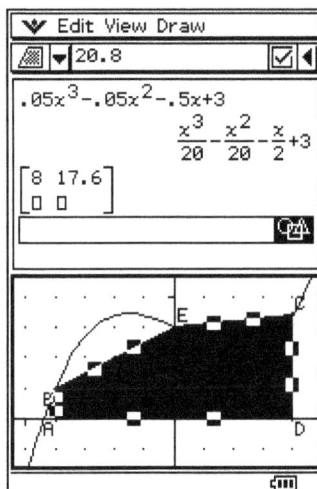

We can repeat the procedure, deleting BE and EC and replacing them with segments BF, FE, EG, GC where F and G are points on the curve at x = -2 and 2 respectively:

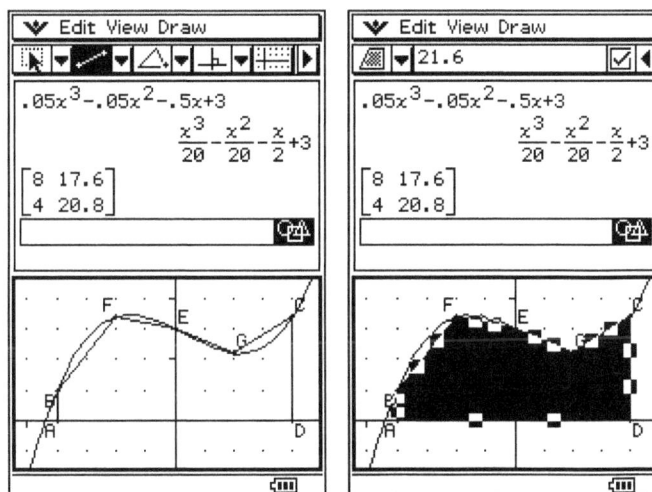

We need an extra row in our matrix to put this data in: we create this using the **Matrix Extend Downwards** button in the 2D Math Keyboard:

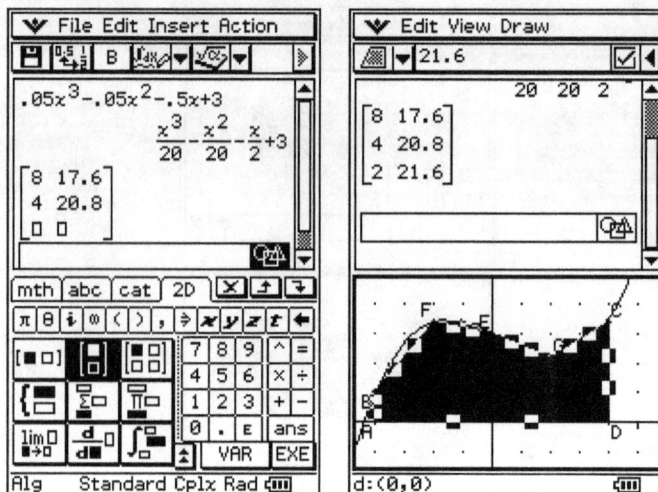

We can superimpose these approximations on the graph by dragging and dropping the matrix: A matrix of 2 columns and 3 or more rows is interpreted as an open polygonal curve. (Piecewise straight line to use other words).

Of course you'll have to zoom out to see the curve as it is up around $y = 20$.

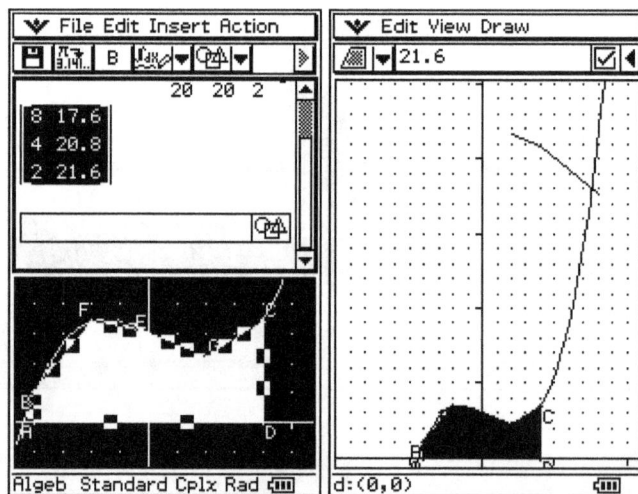

As we keep reducing the length of the line segments which we are using to approximate the curve, the polygon's area will be a closer and closer approximation for the area under the curve. Ideally we would end up with an infinite number of segments each of length 0 and this would give us the exact area (but take a long time to draw!)

Obviously we cannot create an infinite number of segments. We could take one more step and call it good. But before we do, let's see if we could take many more steps at once. Let's see if we can get ClassPad to come up with a formula for the area of the approximate polygon. The ClassPad command to create a general formula from a list of numbers is **sequence()**.

sequence() has two different forms:

sequence($\{y_1, y_2, .., y_n\}$) finds a formula f(x) such that f(1)= $y_1$, f(2)= $y_2$, .. , f(n)= $y_n$.

sequence($\{x_1, x_2, .., x_n\}, \{y_1, y_2, .., y_n\}$) finds a formula f(x) such that f($x_1$)= $y_1$, f($x_2$)= $y_2$, .. , f($x_n$)= $y_n$.

We should use the second form - and don't forget the curly brackets {}:

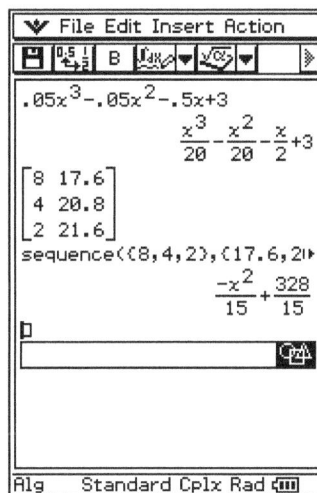

```
▼ File Edit Insert Action
🖫 🔣 B 🔳▼🔳▼    ▶

.05x³-.05x²-.5x+3
                  x³   x²   x
                  ── - ── - ─ +3
                  20   20   2
┌8  17.6⎤
│4  20.8│
└2  21.6⎦
sequence(⟨8,4,2⟩,⟨17.6,2▶
                  -x²   328
                  ─── + ───
                  15    15
□
                              🔳

Alg    Standard Cplx Rad ▥
```

Now let's look at the value this should take at x = 1, and check how close the estimate is with our geometry model:

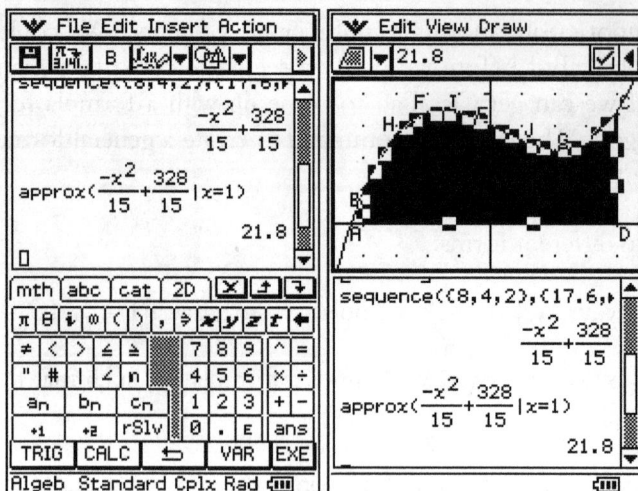

We see that this is in fact spot on. So presumably the formula would be correct for x = 0.5, x= 0.25, etc.

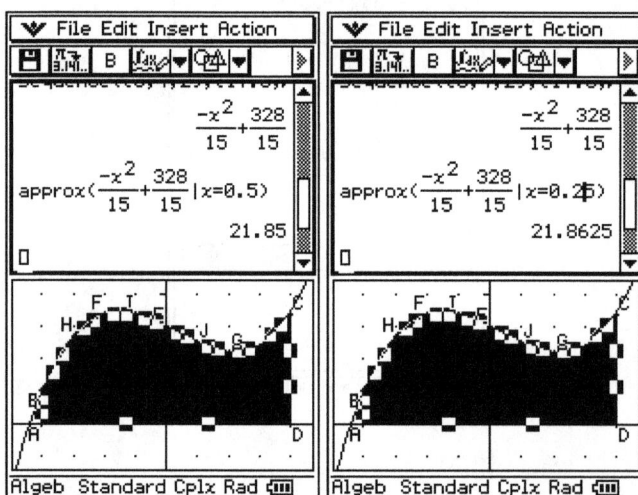

We can put these values into our matrix and drag it into the geometry model to see how it extends the approximation curve:

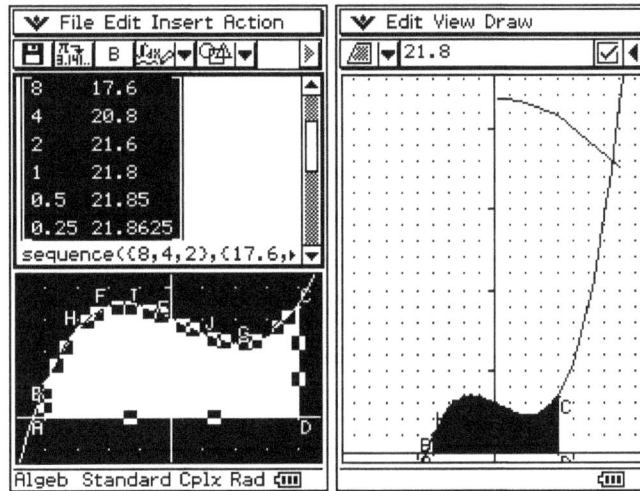

The true area under the curve would be the limit as x goes to zero of this expression, which is just the constant term 328/15

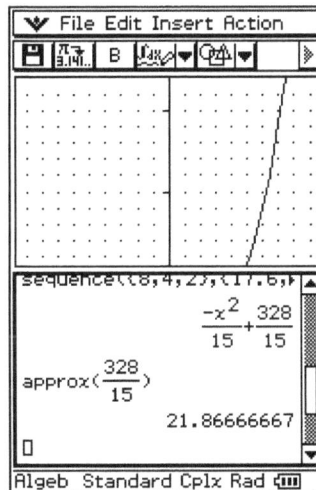

# Areas and Integrals

Let's assume we want to approximate the area under the curve $y=f(x)$ from 0 to x.

As we saw in the previous example, we can estimate this by finding the area of a collection of trapezoids which approximate the curve. Let's remind ourselves of the formula for the area of a trapezoid.

## Area of a Trapezoid

Create a Geometry strip inside your eActivity and draw a trapezoid:

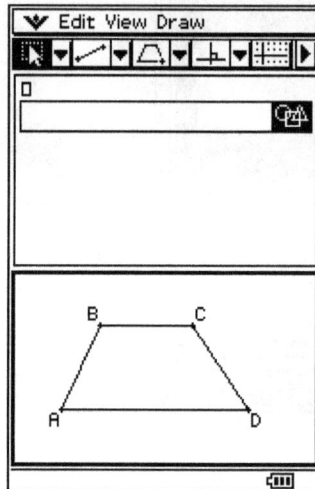

Now measure the length of BC and drag it into a math expression in eActivity: Then measure AD and add its length to BC:

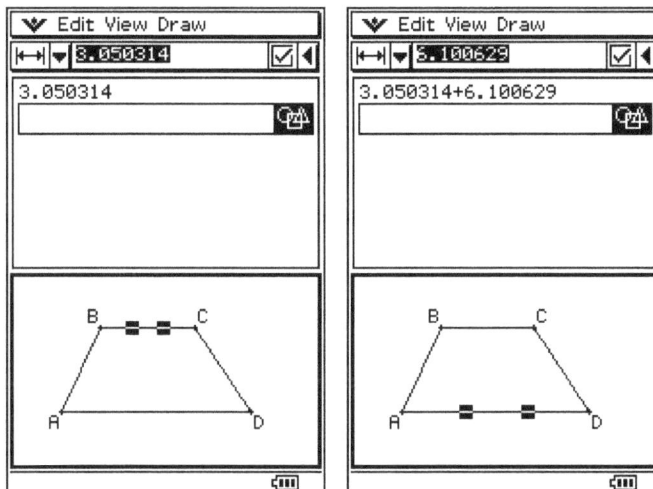

Divide this by two and multiply by the distance between AD and BC. Then select the whole trapezoid and measure its area

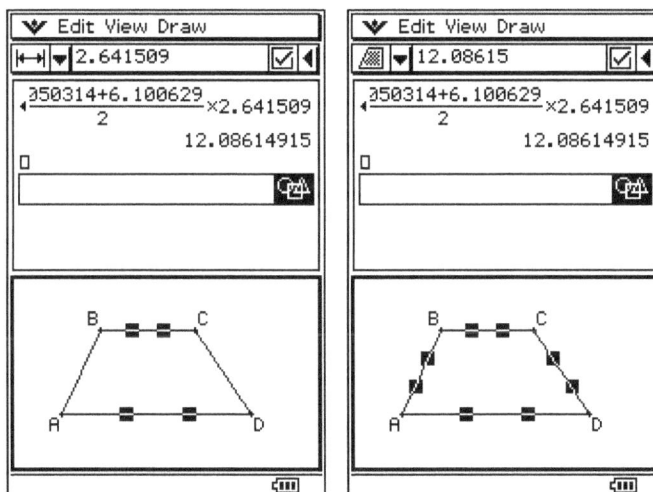

Observe that the trapezoid's area is identical (to within the accuracy of the geometry system) to half the sum of the lengths of the parallel sides times the distance between them .

Now, if our trapezoid is positioned underneath a particular curve, this can be expressed more neatly. Let's examine the curve

$$f(x) = \sin(x)$$

First, create the curve sin(x), then draw a trapezoid so that the points BC lie on the curve. Then constrain the line AD to be the line y=0 and constrain the directions of AB and CD to be vertical:

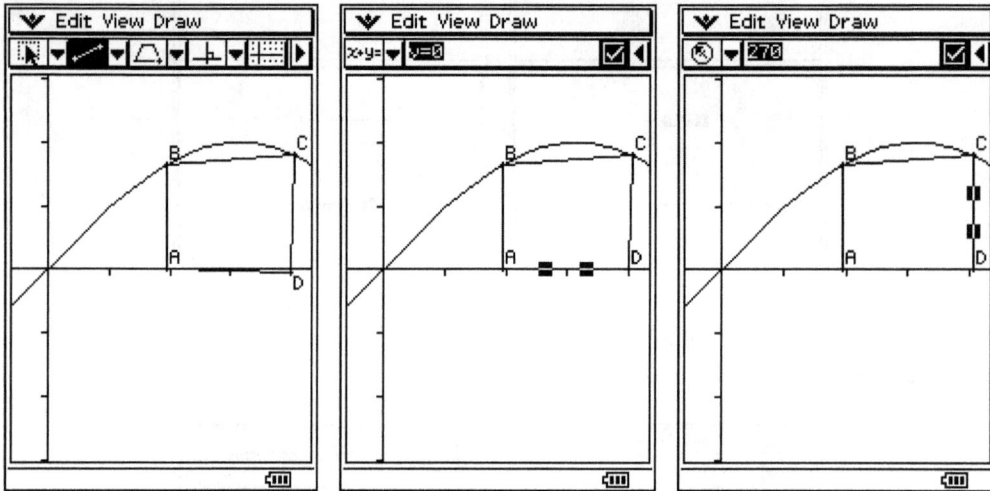

Now slide B so that its x-coordinate is 1, and slide C so that its x-coordinate is 1.5:

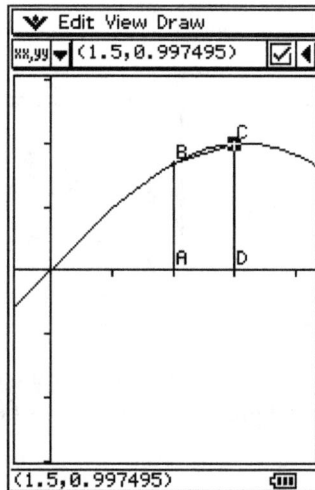

The length of AB is simply the y coordinate of B which is sin(1). And the length of CD is the y coordinate of C, which is sin(1.5). The distance between the parallel sides of the trapezoid is just the difference in x coordinates between A and D, which is 0.5. So we can compute the area of the trapezoid as follows:

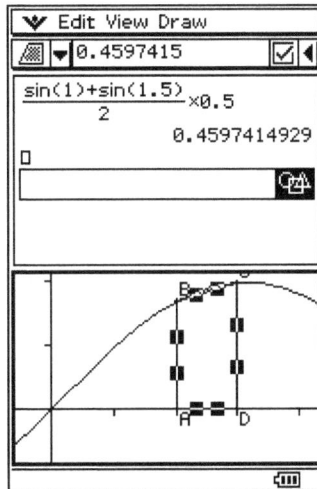

More generally, the area of a trapezoid underneath the curve y=f(x) between the points (x,f(x)) and (x+h,f(x+h)) is:

$$\frac{f(x)+f(x+h)}{2} \cdot h$$

## Area Under a Curve

We can estimate the area under a curve by dividing the area up into a collection of trapezoids, each one of width h. The x coordinate of the left hand side of the ith trapezoid (starting counting at 0) is then ih. The x coordinate of the right hand side of the trapezoid is (i+1)h

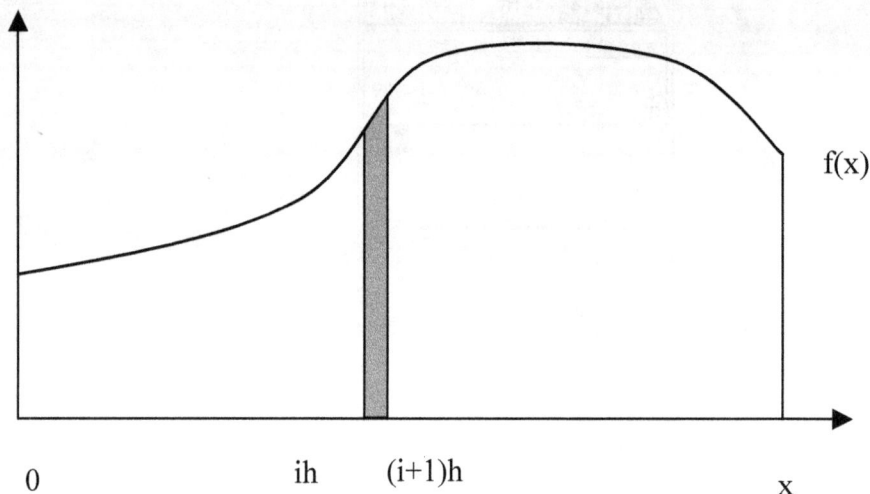

The area of the trapezoid is:

$$\frac{f(i \cdot h) + f((i+1) \cdot h)}{2} \cdot h$$

Hence the trapezoidal approximation to the area under the curve is:

$$\sum_{i=0}^{n-1} \frac{f(i \cdot h) + f((i+1) \cdot h)}{2} \cdot h$$

Where $h = \dfrac{x}{n}$

Let's look at the trapezoidal approximation to the integral for $f(x) = x^2$

```
W File Edit Insert Action

n-1 ⎛   (ih)²+((i+1)h)² ⎞
 Σ  ⎜ h ───────────────── ⎟
i=0 ⎝          2          ⎠

                2·h³·n³+h³·n
                ────────────
                     6
expand(ans)

                 h³·n³   h³·n
                 ───── + ────
                   3      6

□

Alg    Standard Cplx Rad
```

```
W File Edit Insert Action

n-1 ⎛   (ih)²+((i+1)h)² ⎞
 Σ  ⎜ h ───────────────── ⎟
i=0 ⎝          2          ⎠

                2·h³·n³+h³·n
                ────────────
                     6
expand(ans)

                 h³·n³   h³·n
                 ───── + ────
                   3      6
          x
ans|h=────
          n
                  x³     x³
                 ──── + ────
                   3     6·n²

□

Alg    Standard Cplx Rad
```

We can see that as n gets larger this area tends to $\dfrac{x^3}{3}$ . If we can't see it, we can always ask ClassPad to tell us the limit:

```
W File Edit Insert Action

i=0 ⎝           2           ⎠

                2·h³·n³+h³·n
                ────────────
                     6
expand(ans)

                 h³·n³   h³·n
                 ───── + ────
                   3      6
          x
ans|h=────
          n
                  x³     x³
                 ──── + ────
                   3     6·n²

lim (ans)
n→∞
                       x³
                      ────
                       3

□

Alg    Standard Cplx Rad
```

Next, let's look at the area under the curve $f(x) = x^3$, but before we do that, try differentiating the result of the previous calculation:

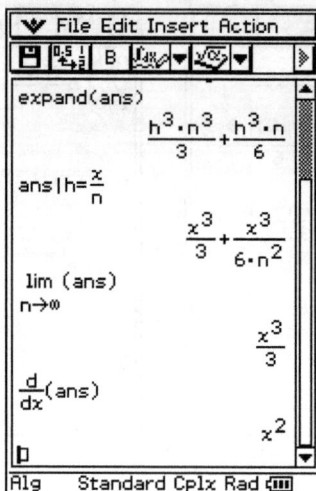

We'll come back to this, but we can repeat our analysis for $f(x) = x^3$ simply by changing the index at the top of the screen:

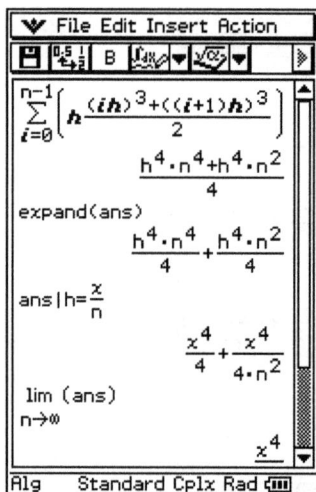

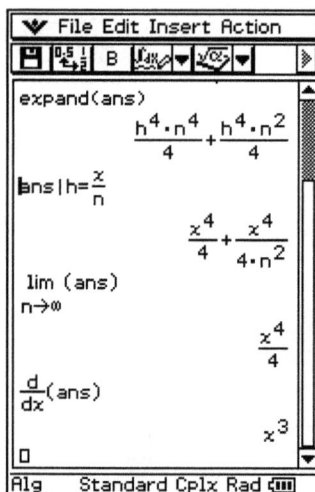

We call the area under the curve the **definite integral** of the curve, and write it like this:

$$\int_{x_0}^{x_1} f(x)\,dx$$

$x_0$ and $x_1$ are called the **limits of integration**.

Of course, ClassPad can compute integrals directly:

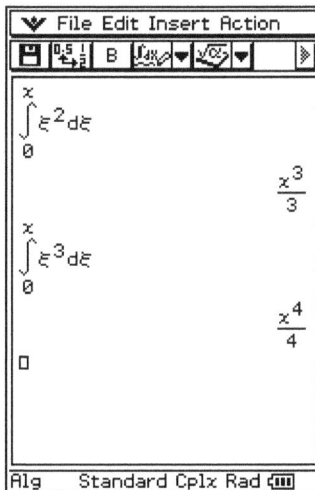

```
┌─────────────────────────────────────┐
│ ❦ File Edit Insert Action           │
├─────────────────────────────────────┤
│ 💾 📊 B 📈▼ 📉▼            ≫        │
├─────────────────────────────────────┤
│  x                                   │
│  ⌠ ε²dε                              │
│  ⌡                                   │
│  0                                   │
│                              x³      │
│                              ──      │
│                              3       │
│  x                                   │
│  ⌠ ε³dε                              │
│  ⌡                                   │
│  0                                   │
│                              x⁴      │
│                              ──      │
│                              4       │
│  □                                   │
│                                      │
│                                      │
├─────────────────────────────────────┤
│ Alg    Standard Cplx Rad ▥          │
└─────────────────────────────────────┘
```

EXERCISES

Use the methods of this chapter to derive the area under the following curves:

1. $f(x) = (x-2)^2$

2. $f(x) = e^x$

3. Differentiate the solutions to 1 and 2.

## Fundamental Theorem of Calculus

For the functions $f(x) = x^3$ and $f(x) = x^2$ we saw that the derivative of the integral of f was in fact f. That is:

$$\frac{d}{dx} \int_0^x f(\xi) d\xi = f(x)$$

Is this generally true? Try a few examples in ClassPad to start convincing yourself.

| ▼ File Edit Insert Action | ▼ File Edit Insert Action | ▼ File Edit Insert Action |
|---|---|---|
| $\int_0^x \cos(\xi) d\xi$ | $\int_0^x \tan(\xi) d\xi$ | $\int_0^x \ln(\xi) d\xi$ |
| $\sin(x)$ | $-\ln(\cos(x))$ | $x \cdot \ln(x) - x$ |
| $\frac{d}{dx} \text{ans}$ | $\frac{d}{dx} \text{ans}$ | $\frac{d}{dx} \text{ans}$ |
| $\cos(x)$ | $\dfrac{\sin(x)}{\cos(x)}$ | $\ln(x)$ |
| □ | □ | □ |
| Alg   Standard Cplx Rad | Alg   Standard Cplx Rad | Alg   Standard Cplx Rad |

This is a very important observation.  So important it is given the name of the Fundamental Theorem of Calculus. We'd better examine it a little.

## Graphical Analysis

Let's first look at the theorem in terms of graphs in a geometry window. Create the following graph:

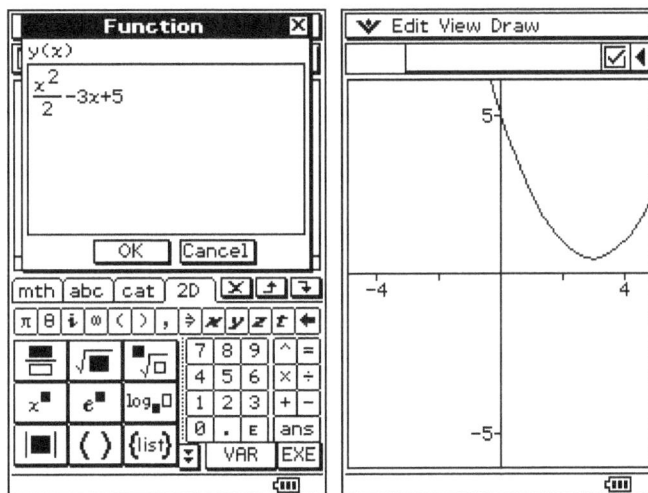

We'll use the Animation capabilities of ClassPad to examine the trapezoidal approximation to the integral.

First we'll create a point A on the curve at x=0. Then we create a second point B on the curve, select B and the curve and Edit/Animate/Add Animation:

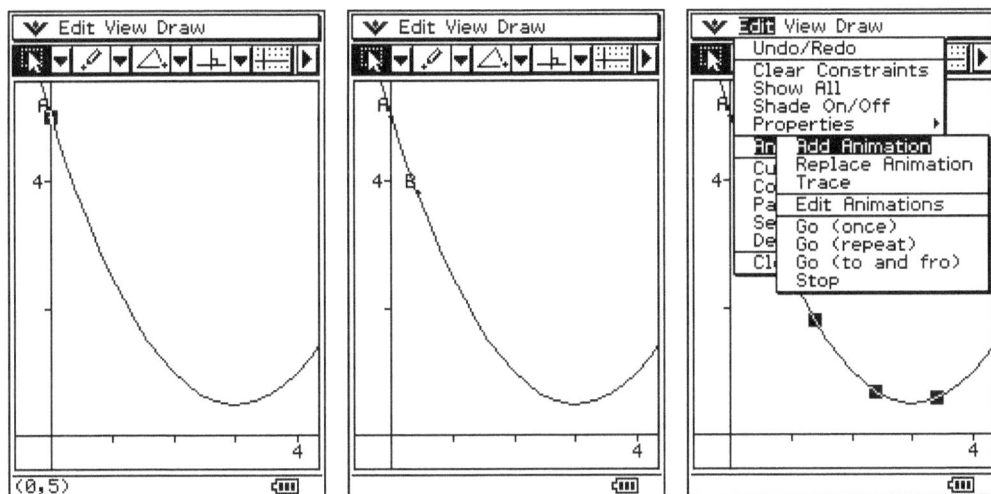

Now, we'll start B at the point with x=0.2 and end it at the point with x=1. To do this we go int the Edit Animations window:

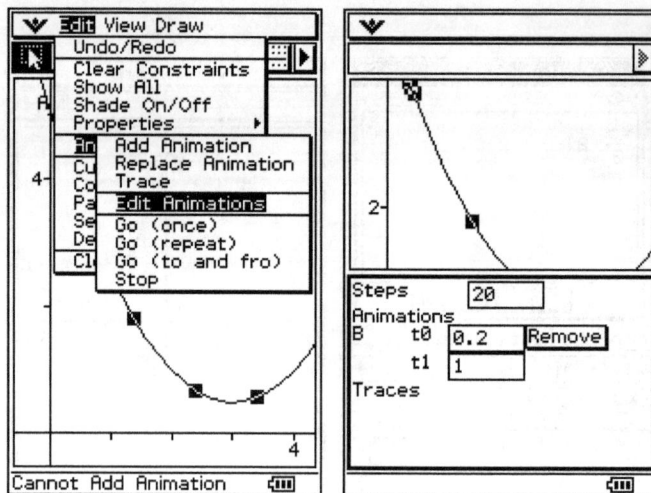

Try running the animation (select View / Animation UI): you'll see B move along the curve as specified.

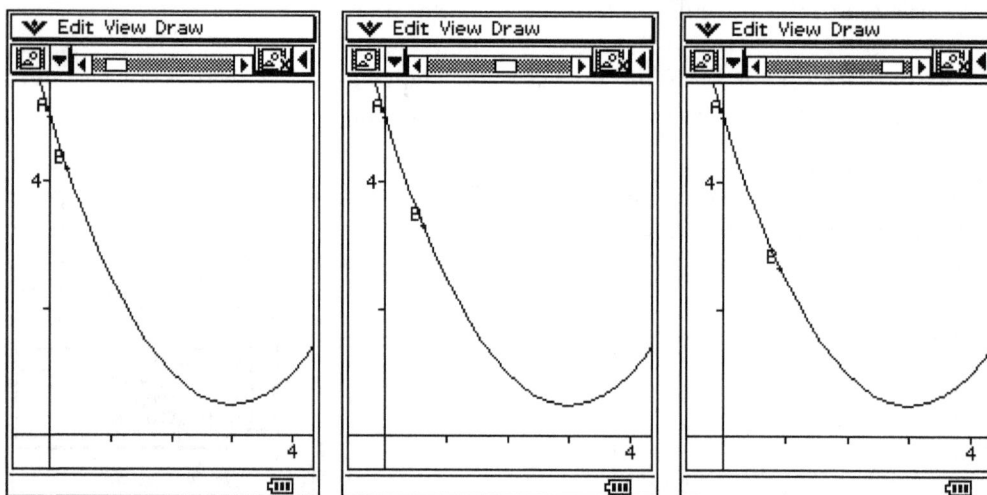

We now want a third point C, which we will run from 0.4 to 2:

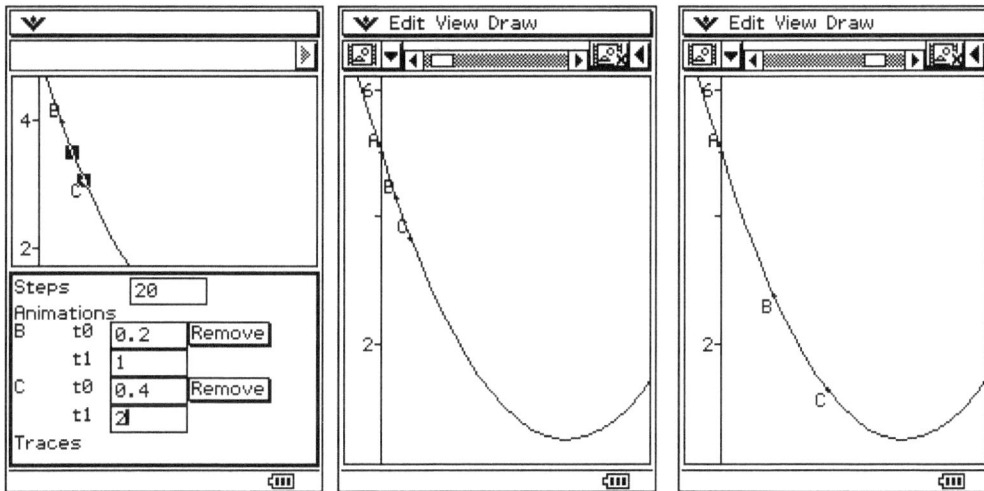

We now add 3 more points D, E and F, setting their animations appropriately:

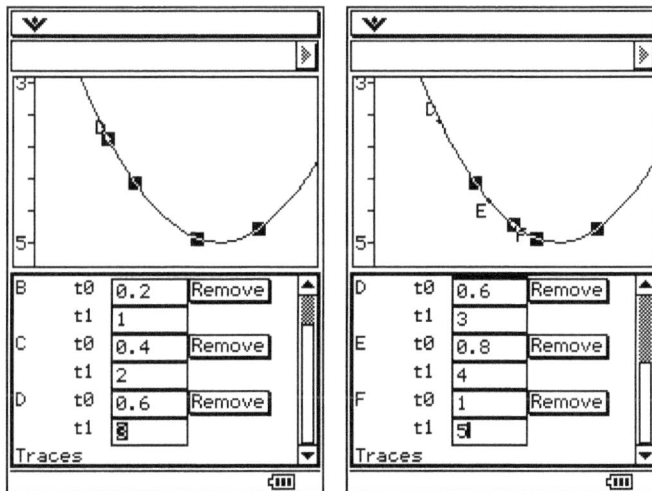

Our next step is to join up the dots and construct the polygon whose area approximates the area under the curve. Hide the curve and the axes to make it easier to see:

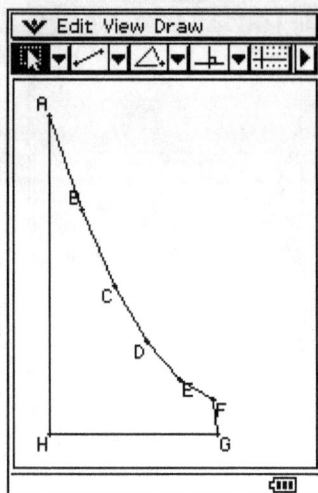

We need to make sure that the line FG is vertical. We can do this by setting its direction. While we are at it we can set the direction of AH and the equation of GH to make sure they remain the axes:

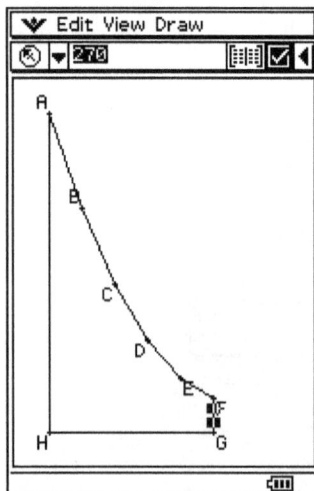

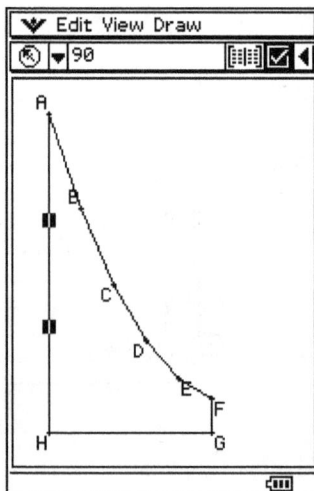

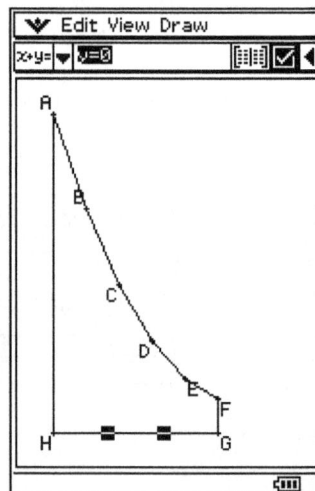

We can now shade this polygon and animate:

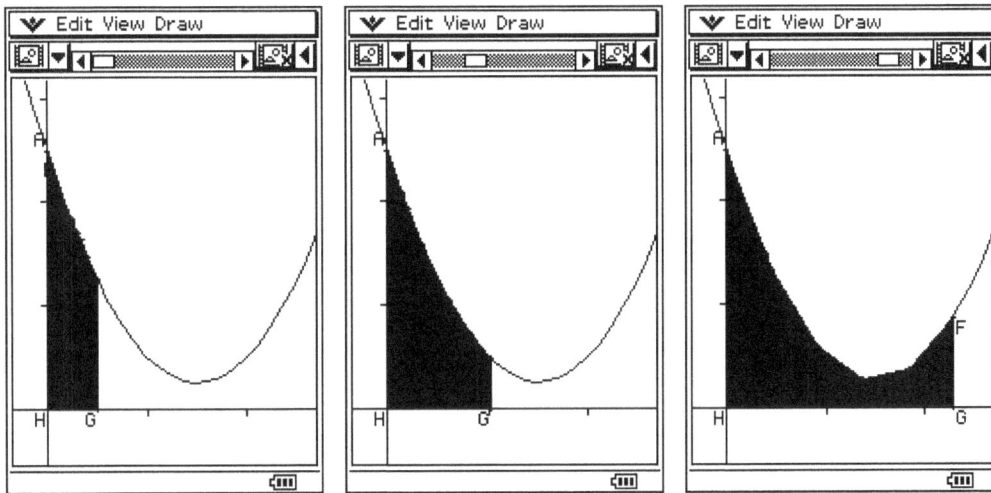

We can look at the area of the shaded polygon simply by leaving nothing selected and bringing up the measure box (by default it displays the area of the shaded polygon). Pressing the Table button will tabulate the area values:

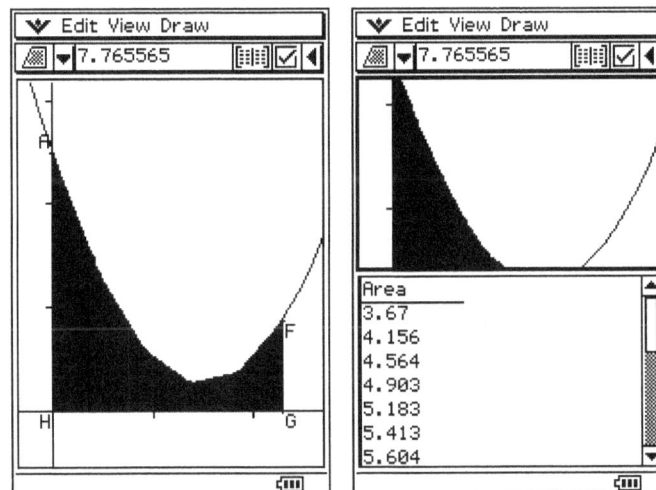

Tabulate the location of point G by selecting it and pressing the Table button. You can then use Edit/Move to Front to put the x coordinates in front of the Area.

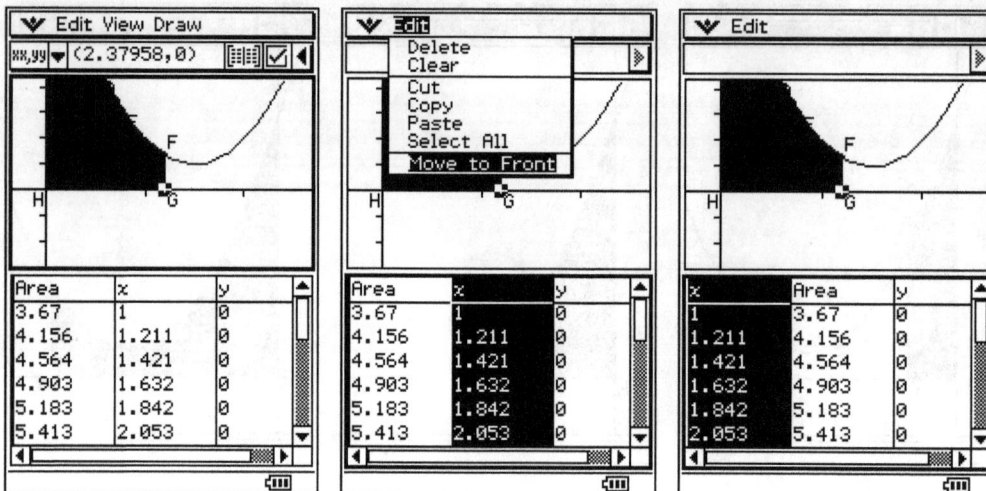

Dragging both columns of the table back into Geometry gives a graph of the area of the polygon:

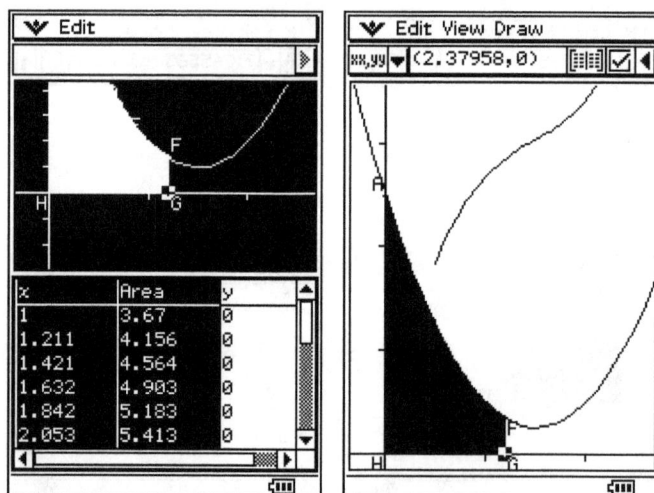

Next step is to run a point along the new curve. Create the point I on the curve, and then add an animation. We see that I runs along the area curve as our polygon expands:

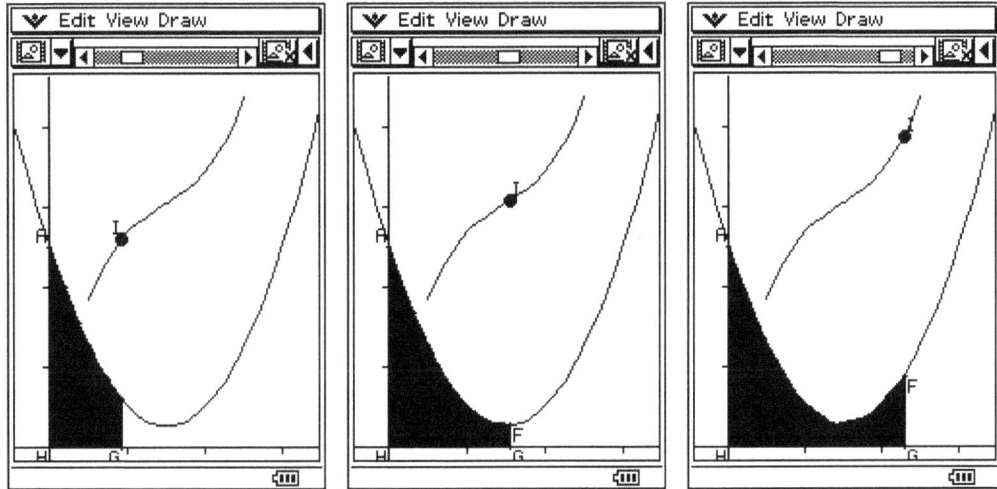

Next step is to add another point J to run along the area curve. Initially, the animations of I and J are both set to run along the curve from parameter value 0 to parameter value 19 (these parameter values refer to the 20 points captured in our table.) We reset the end values so that I runs from 0 to 18 and J runs from 1 to 19:

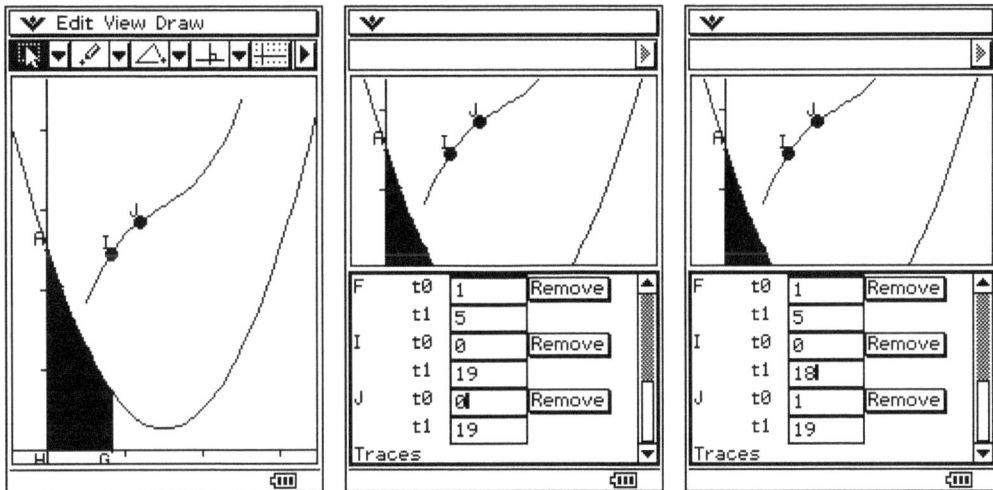

Create a line through I and J. This is a chord to the curve.

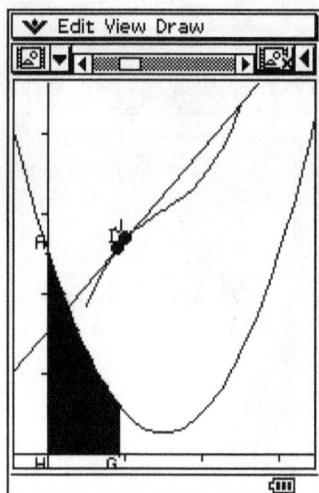

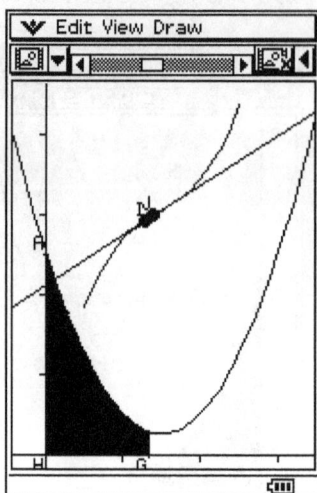

Finally, tabulate the slope of the chord against the x coordinate of I:

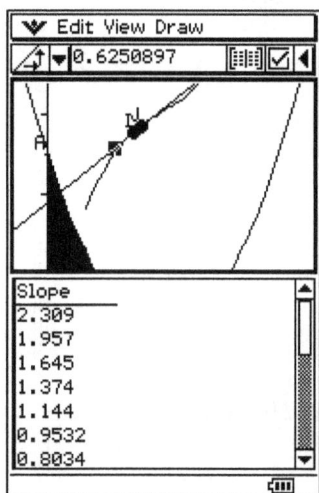

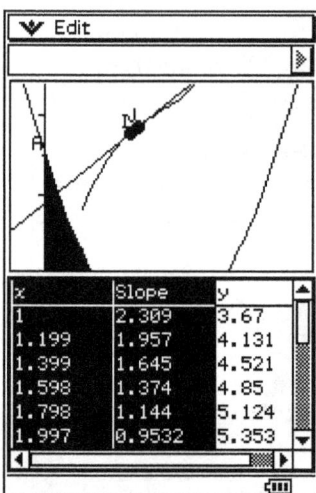

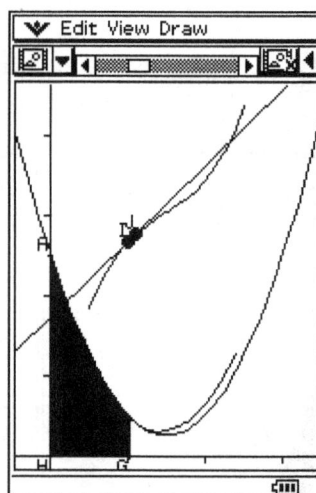

| x | Slope | y |
|---|-------|---|
| 1 | 2.309 | 3.67 |
| 1.199 | 1.957 | 4.131 |
| 1.399 | 1.645 | 4.521 |
| 1.598 | 1.374 | 4.85 |
| 1.798 | 1.144 | 5.124 |
| 1.997 | 0.9532 | 5.353 |

Hiding the polygon away, we can observe that the resultant curve is pretty close to the original.

## Algebraic Analysis

Let's examine the trapezoidal area / chord slope relationship in a purely algebraic context. Given a function f(x), let s(x) be the slope of the chord starting at x and ending at x+h:

Now let's approximate the area under the curve s(x) using trapezoids of width h.

We write t(i) for the i$^{th}$ trapezoid and add up 3 of them to start with:

```
▼ File Edit Insert Action                ≫
💾 📊 B 📈▼ 📖▼              ≫

define s(x)= f(x+h)-f(x)
              ─────────
                  h
                                    done
◀he t(i)=h× s(i*h)+s((i+1)h)
            ───────────────
                   2
                                    done
expand(t(0)+t(1)+t(2))
 -f(h)   f(4·h)   f(3·h)   f(0)
 ───── + ────── + ────── - ────
   2       2        2       2
▯

Alg    Standard Cplx Rad ▥
```

We notice that the formula is not quite as complicated as it could be as it has no term which involves f(2h).  Big deal, we got 4 terms instead of 5.   But the saving is more pronounced if we split the problem up into 4 trapezoids, or 5, or 6 (assuming that as we add trapezoids, we decrease the size of h in such a way that with n trapezoids, n.h = x : i.e. h = x/n)

```
▼ File Edit Insert Action       ≫
💾 📊 B 📈▼ 📖▼        ≫

define s(x)= f(x+h)-f(x)
              ─────────
                  h
                            done
◀he t(i)=h× s(i*h)+s((i+1)h)
            ───────────────
                   2
                            done
◀▷and(t(0)+t(1)+t(2)+t(3))
 -f(h)  f(5·h)  f(4·h)  f(0)
 ───── +────── +────── -────
   2      2       2      2
▯




Alg   Standard Cplx Rad ▥
```

```
▼ File Edit Insert Action       ≫
💾 📊 B 📈▼ 📖▼        ≫

define s(x)= f(x+h)-f(x)
              ─────────
                  h
                            done
◀he t(i)=h× s(i*h)+s((i+1)h)
            ───────────────
                   2
                            done
◀t(0)+t(1)+t(2)+t(3)+t(4)▷
 -f(h)  f(6·h)  f(5·h)  f(0)
 ───── +────── +────── -────
   2      2       2      2
▯




Alg   Standard Cplx Rad ▥
```

```
▼ File Edit Insert Action       ≫
💾 📊 B 📈▼ 📖▼        ≫

define s(x)= f(x+h)-f(x)
              ─────────
                  h
                            done
◀he t(i)=h× s(i*h)+s((i+1)h)
            ───────────────
                   2
                            done
◀t(1)+t(2)+t(3)+t(4)+t(5))
 -f(h)  f(7·h)  f(6·h)  f(0)
 ───── +────── +────── -────
   2      2       2      2
▯




Alg   Standard Cplx Rad ▥
```

All the middle terms disappear, and we notice we can rearrange the terms as follows:

```
 ‾W‾ File Edit Insert Action
 ┌──┬──┬─┬──┬─┬──┬─┐        ┌─┐
 │⊟ │⊞ │B│A│▼│✎│▼│        │»│
 └──┴──┴─┴──┴─┴──┴─┘        └─┘
 ┌──────────────────────────────┐
 │                 f(x+h)-f(x)   │
 │ define s(x)=───────────       │
 │                    h          │
 │                         done  │
 │            s(i*h)+s((i+1)h)   │
 │ ◀ne t(i)=h×───────────────    │
 │                  2            │
 │                         done  │
 │ ◀t(1)+t(2)+t(3)+t(4)+t(5))    │
 │  -f(h)   f(7·h)   f(6·h)  f(0)│
 │  ───── + ────── + ────── ─────│
 │    2       2        2      2  │
 │                               │
 │  f(6·h)+f(7·h)   f(h)+f(0)    │
 │  ───────────── ─ ─────────    │
 │        2             2        │
 │ ▯                             │
 │ ❘                             │
 │                               │
 └──────────────────────────────┘
 ┌────────────────────────────────┐
 │Alg    Standard Cplx Rad ▭▭│
 └────────────────────────────────┘
```

As 6h=x, this can be rewritten:

$$\frac{f(x)+f(x+h)}{2} - \frac{f(0)+f(h)}{2}$$

The above analysis was all algebraic, and nowhere involved limits. The important point was that, because of the close relationship between the formula for the slope of a chord and for the area of a trapezoid of the same width, most of the terms in the trapezoidal approximation for the area cancelled out.

It is suggestive to note that as h becomes small, we can expect

$$f(x+h) \rightarrow f(x)$$

and

$$f(h) \rightarrow f(0)$$

hence

$$\frac{f(x)+f(x+h)}{2} - \frac{f(0)+f(h)}{2} \rightarrow f(x) - f(0)$$

EXERCISES

1.  The graphical analysis in the first part of this chapter looked first at area then at chord slope.  Do the graphical analysis the other way round:  first creating a graph of chord slope, then finding the area of trapezoids subtended by the chord slope graph.

2.  The algebraic analysis of the second part of the chapter found the chord slope first, then added the areas of trapezoids subtended by the chord slope.  Do this analysis the other way round (area first, then chord slope).

## Anti-Derivatives

The Fundamental Theorem of Calculus says that integration is the inverse operation to differentiation. So the process of integrating can be viewed as the process of anti-differentiating. That is, given a function g(x) that you want to integrate, you need to think of a function f(x) such that:

$$\frac{d}{dx} f(x) = g(x)$$

We write the anti-derivative using the integral symbol , but without filling in the limits of integration:

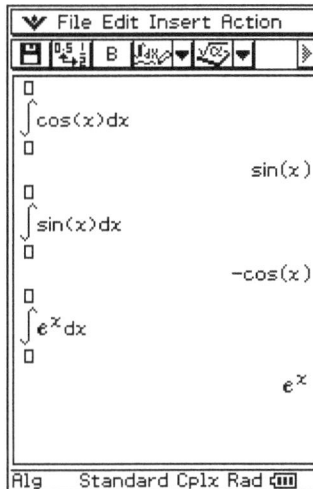

```
▼ File Edit Insert Action
┌────────────────────────────┐
│ □                          │
│ ∫ cos(x)dx                 │
│ □                          │
│                    sin(x)  │
│ □                          │
│ ∫ sin(x)dx                 │
│ □                          │
│                   -cos(x)  │
│ □                          │
│ ∫ eˣdx                     │
│ □                          │
│                       eˣ   │
│                            │
└────────────────────────────┘
Alg    Standard Cplx Rad 📶
```

Unfortunately, in general, finding an anti-derivative is considerably more difficult than just differentiating a known function f(x).

However it's easy enough for some simple functions. For example, let's find the anti-derivative of $x^n$. A little trial and error gets us there. Of course the indefinite integral agrees with us:

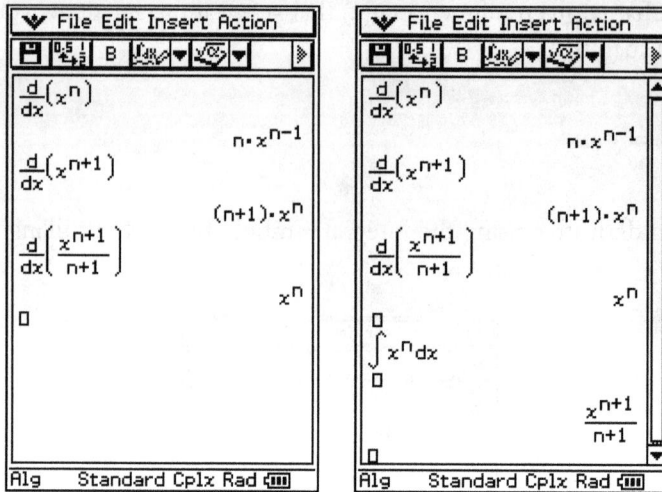

| File Edit Insert Action |
|---|
| $\frac{d}{dx}\left(x^n\right)$ |
| $n \cdot x^{n-1}$ |
| $\frac{d}{dx}\left(x^{n+1}\right)$ |
| $(n+1) \cdot x^n$ |
| $\frac{d}{dx}\left(\frac{x^{n+1}}{n+1}\right)$ |
| $x^n$ |
| □ |
| Alg    Standard Cplx Rad |

| File Edit Insert Action |
|---|
| $\frac{d}{dx}\left(x^n\right)$ |
| $n \cdot x^{n-1}$ |
| $\frac{d}{dx}\left(x^{n+1}\right)$ |
| $(n+1) \cdot x^n$ |
| $\frac{d}{dx}\left(\frac{x^{n+1}}{n+1}\right)$ |
| $x^n$ |
| □ $\int x^n dx$ □ |
| $\frac{x^{n+1}}{n+1}$ |
| □ |
| Alg    Standard Cplx Rad |

Use the same technique to find the anti-derivative of

$\cos(3x)$

$e^{\frac{x}{2}}$

$\sin(Ax)$

In Each case, we start with a guess then adjust it.

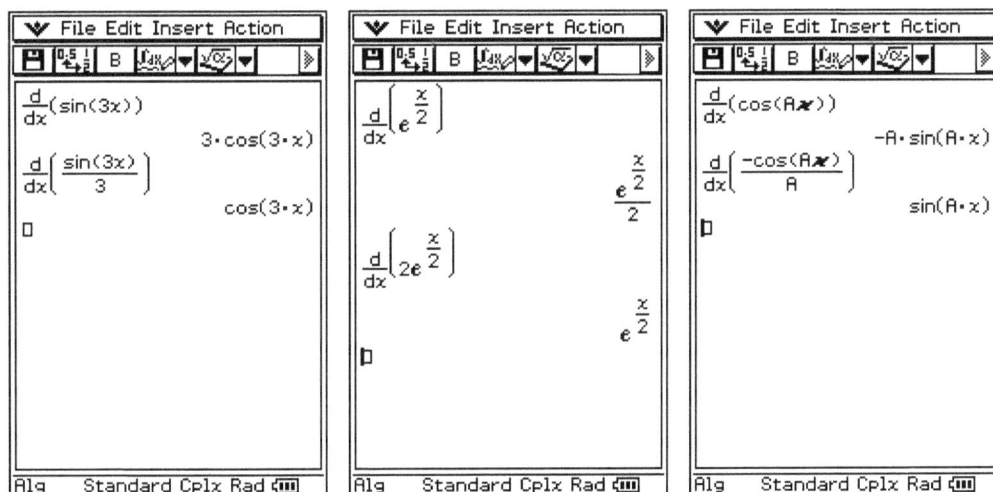

$$\frac{d}{dx}(\sin(3x))$$
$$3 \cdot \cos(3 \cdot x)$$
$$\frac{d}{dx}\left(\frac{\sin(3x)}{3}\right)$$
$$\cos(3 \cdot x)$$

$$\frac{d}{dx}\left(e^{\frac{x}{2}}\right)$$
$$\frac{e^{\frac{x}{2}}}{2}$$
$$\frac{d}{dx}\left(2e^{\frac{x}{2}}\right)$$
$$e^{\frac{x}{2}}$$

$$\frac{d}{dx}(\cos(Ax))$$
$$-A \cdot \sin(A \cdot x)$$
$$\frac{d}{dx}\left(\frac{-\cos(Ax)}{A}\right)$$
$$\sin(A \cdot x)$$

In general, we see that this sort of example comes from the chain rule:

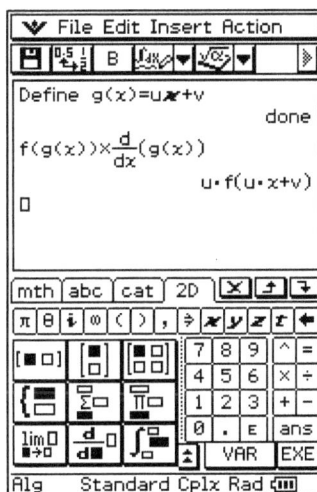

$$\text{Define } g(x) = ux + v$$
$$\text{done}$$
$$f(g(x)) \times \frac{d}{dx}(g(x))$$
$$u \cdot f(u \cdot x + v)$$

EXERCISES

Guess anti-derivatives to the following functions. Check your results by differentiating in ClassPad:

1. $e^{-kx}$

2. $2\sin(x)$

3. $\dfrac{1}{x}$

# Integrate Af(x)+Bg(x)

What happens to the integral when we multiply a function by a scalar, or when we add two functions?

## Integrate Scalar Multiple

First let's look at a graph of the function:

$$y(x) = x^2 + 1$$

Zoom in on it appropriately and draw the trapezoid between 0 and 1 (turning on the integer grid helps in the accurate placement of the points). Shade the trapezoid. With nothing selected, the measure box gives the shaded area.

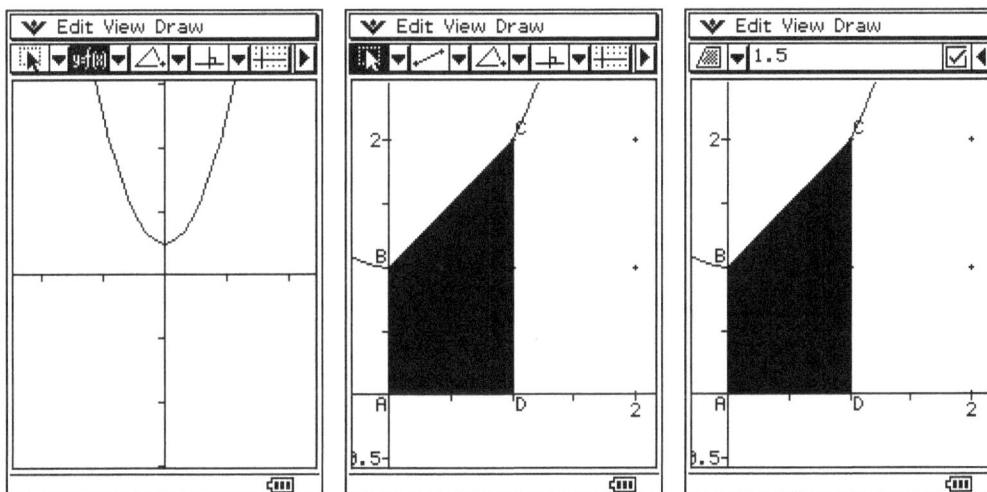

Change the function to

$$y(x) = \frac{x^2 + 1}{2}$$

Observe how the area changes:

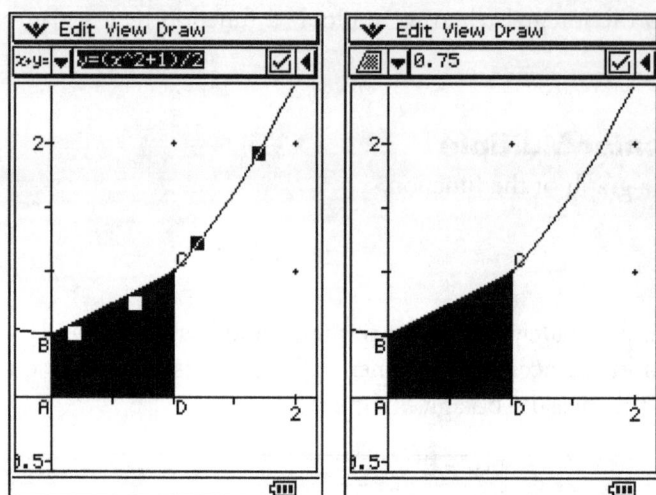

Now change the function to

$$y(x) = 2(x^2 + 1)$$

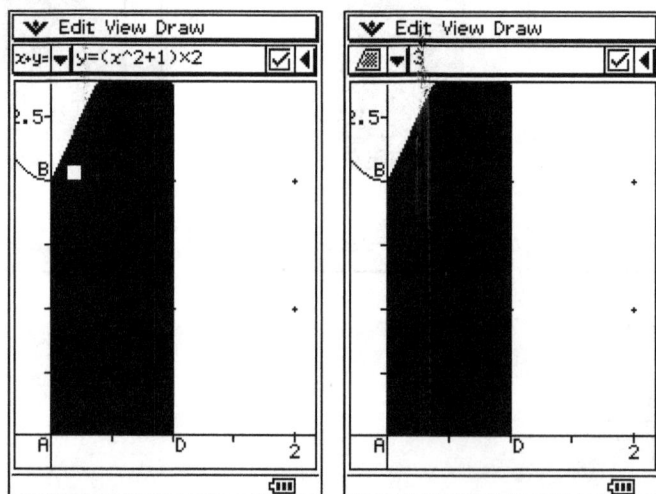

We see that the area of the trapezoid is proportional to the multiplier of the function.

Let u(x) be the anti-derivative of f(x). If we multiply u(x) by a constant A then differentiate, we see that the result is A.f(x):

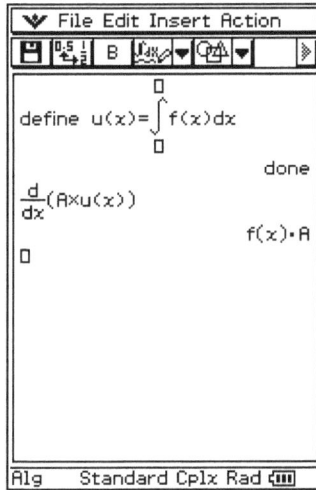

Hence the following:

$$\int Af(x)dx = A \int f(x)dx$$

## Integrate Sum

Now let's consider the sum of two functions. First, in geometry revert to the original y(x), and copy the trapezoid area into the eActivity window:

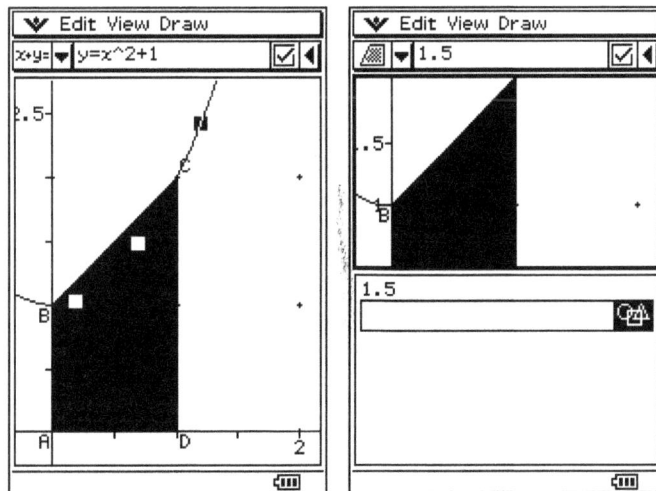

Now change the function to cos(x) and copy the area, adding it to the area of the original trapezoid:

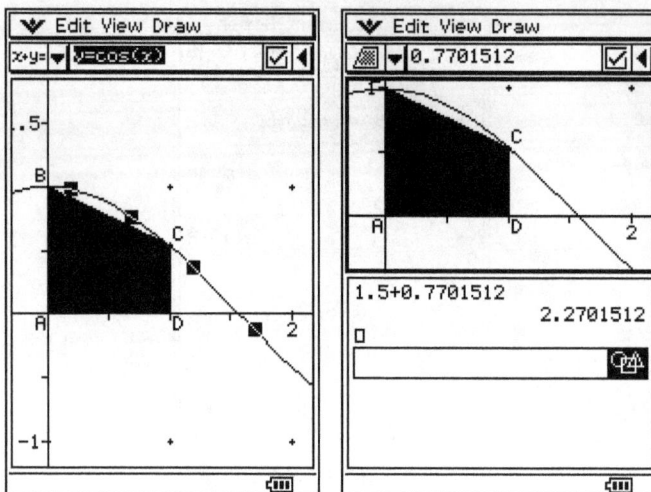

Now change the function to

$$y(x) = x^2 + 1 + \cos(x)$$

Observe that the area of the new trapezoid is the sum of the areas of the previous two trapezoids:

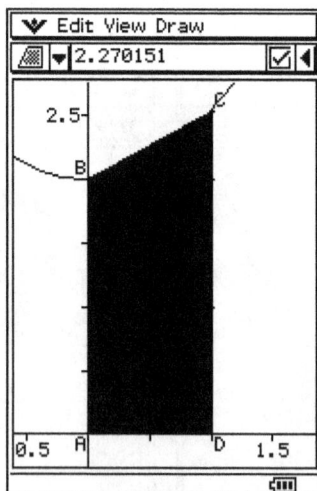

Let u(x) be the anti-derivative of f(x) and let v(x) be the anti-derivative of g(x). We can see that the derivative of u(x)+v(x) = f(x)+g(x):

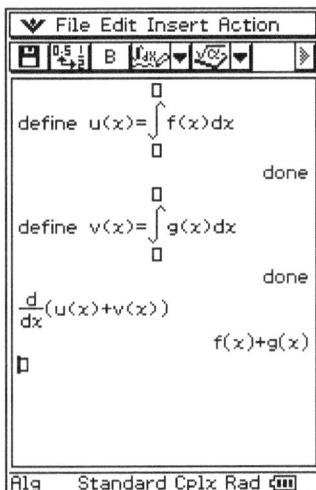

```
▼ File Edit Insert Action
[icons]                              ≫
                 □
define u(x)=∫ f(x)dx
                 □
                                 done
                 □
define v(x)=∫ g(x)dx
                 □
                                 done
 d
 ── (u(x)+v(x))
 dx
                         f(x)+g(x)
□

Alg    Standard Cplx Rad (▥)
```

Hence u+v is the anti-derivative of f+g. That is:

$$\int f(x) + g(x)dx = \int f(x)dx + \int g(x)dx$$

## Integrate Linear Combination

These two results can be combined in the following one:

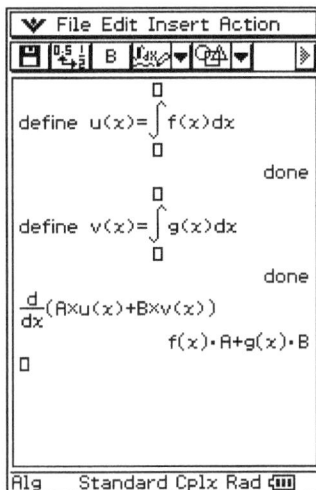

```
▼ File Edit Insert Action
[icons]                              ≫
                 □
define u(x)=∫ f(x)dx
                 □
                                 done
                 □
define v(x)=∫ g(x)dx
                 □
                                 done
 d
 ── (A×u(x)+B×v(x))
 dx
                      f(x)·A+g(x)·B
□

Alg    Standard Cplx Rad (▥)
```

Hence this:

$$\int Af(x) + Bg(x)dx = A\int f(x)dx + B\int g(x)dx$$

EXERCISES

Do the following integrations.  Check your answers with ClassPad

1.  $\int e^x + 2xdx$

2.  $\int 2\sin x - 3\cos xdx$

3.  $\int x^2 + 2x - 3dx$

4.  $\int e^{x+1} + \sin xdx$

# Integrate f(ux+v)

We recall the chain rule for differentiation, which says:

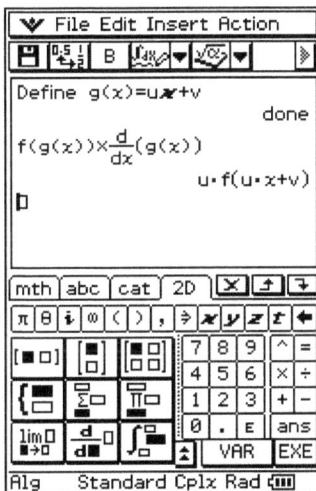

```
▼ File Edit Insert Action
[icons]                              ⟫

Define g(x)=u·x+v
                              done

f(g(x))×d/dx(g(x))
                      u·f(u·x+v)
□

mth  abc  cat  2D  [X][↑][↓]
π θ i ω ( ) , ⇥ x y z t ←
[■□]  [■/□] [■□/□□]  7 8 9  ^ =
                     4 5 6  × ÷
{■/□} Σ□ π□           1 2 3  + −
lim□  d□  ∫□          0 . E  ans
■→□   d■  ∫□   ▲  VAR  EXE
Alg    Standard Cplx Rad ▦
```

From the above screen, we can derive the following:

$$\int f(a.x + b)\,dx = \frac{1}{a}\int f(x)\,dx$$

Let's look at an example of this in a Geometry Window. We will look at a graph of the function

$$y(x) = \sin(2 \cdot x + \frac{3}{2})$$

Go ahead and create this function, along with the function

$$y(x) = \sin(x)$$

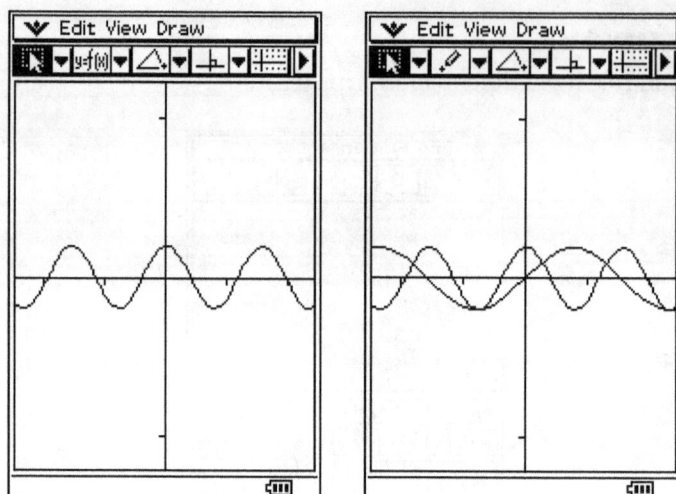

You should also create a line and set its equation to be

$$y = 2 \cdot x + \frac{3}{2}$$

It is important for the rest of the example that you create this using the infinite line tool, rather than creating a function and defining it to have a linear equation. This is because ClassPad is able to find the intersection between two lines, but is not able to find the intersection between a line and a function.

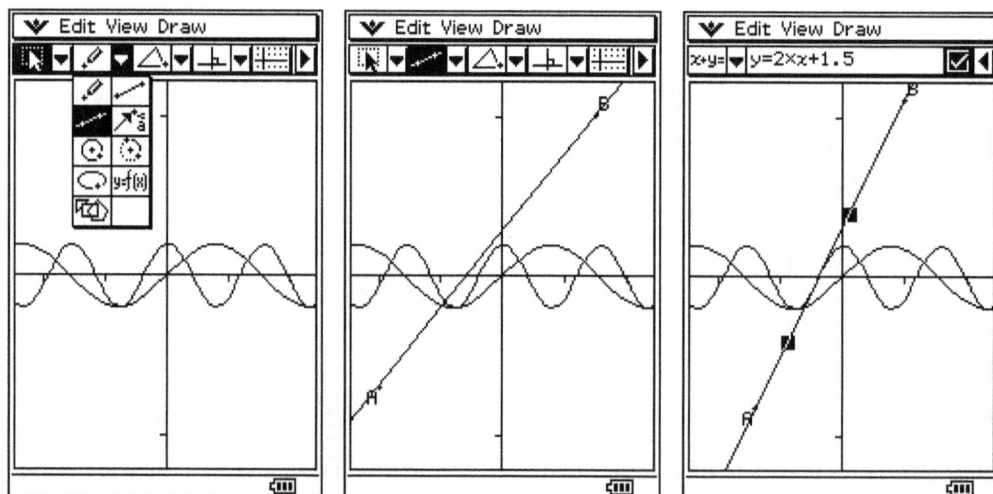

We are going to create two points on our original curve which are going to define the vertices of a trapezoid.

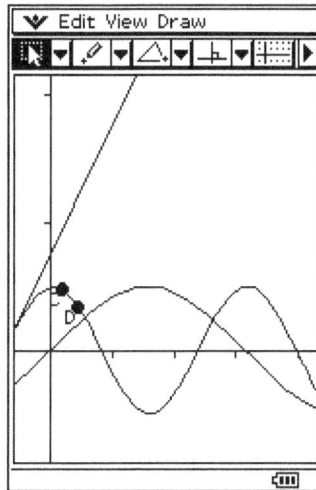

Now, draw a line up from C to AB, and a line from D to AB. Select these lines and constrain them to be vertical:

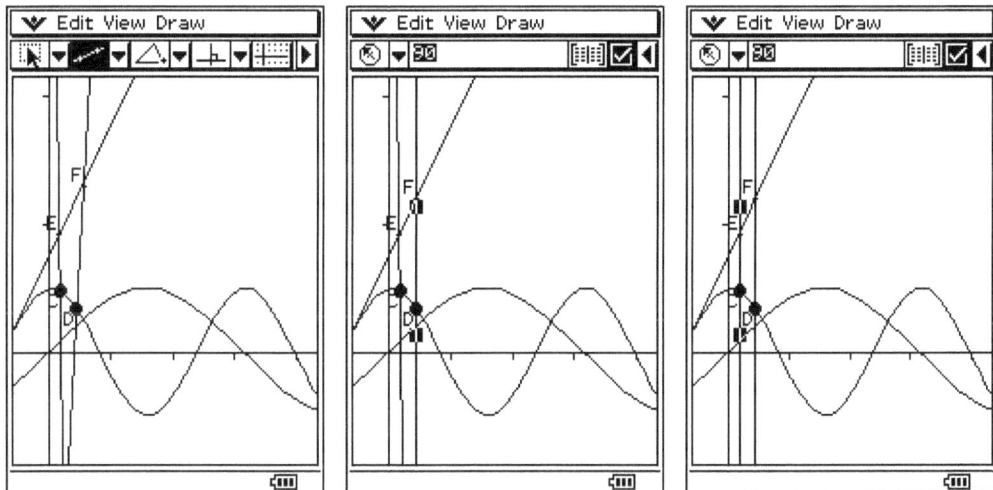

Draw another line GH and set its equation to:    y=x (Thickening the original line and curves and the points C,D,E,F will perhaps ease the confusion in the diagram).

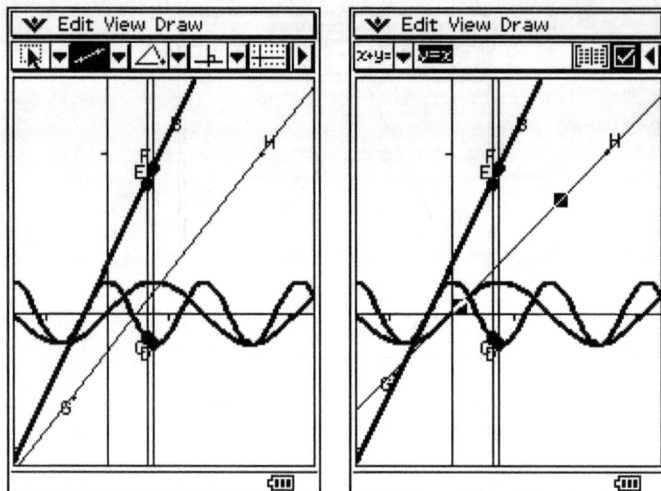

We want to draw lines from E and F to GH, then constrain them to be horizontal:

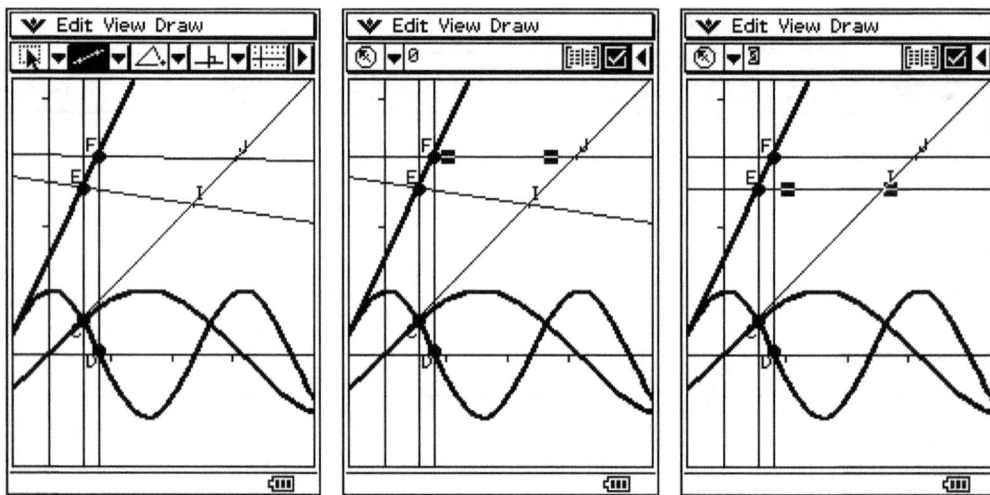

Select the two lines EI, FJ  and reflect them in y=x:

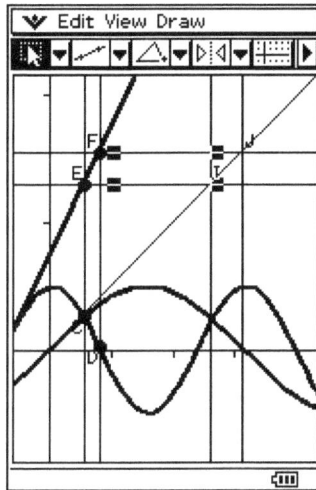

Finally, create lines from C and D to these new lines and constrain them to be horizontal:

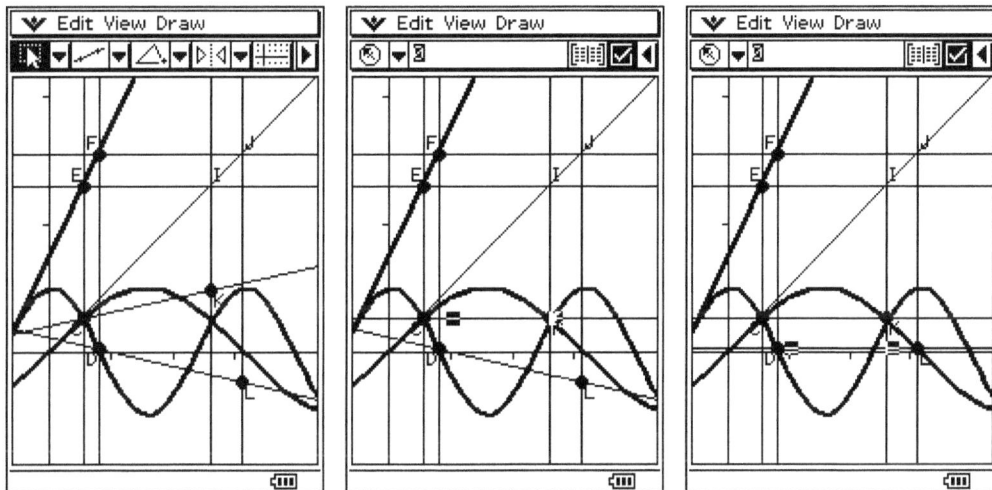

Notice that these new points lie on the y=sin(x) curve, and they stay there as you drag the original points CD around:

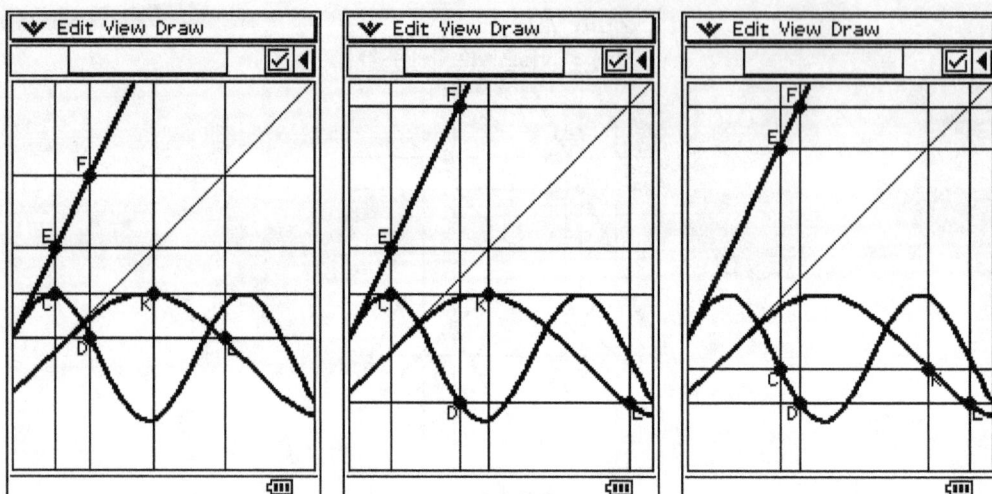

You can examine the horizontal distance between C and D by selecting the two vertical lines and putting distance in the Measure Box. Likewise for K and L:

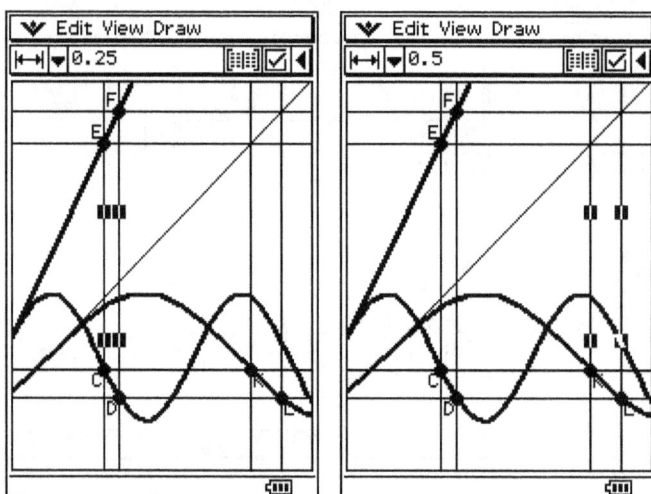

We see that the ratio of the distances is equal to the slope of the line AB:

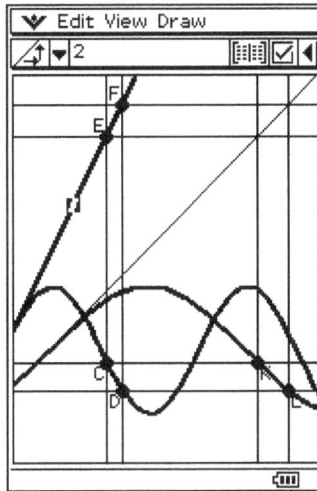

Check this out in a second example by changing the equation of AB, and making the equivalent change to the original curve equation:

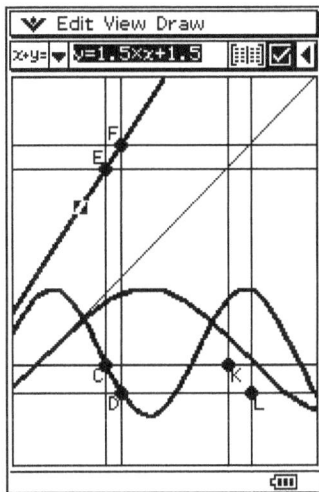

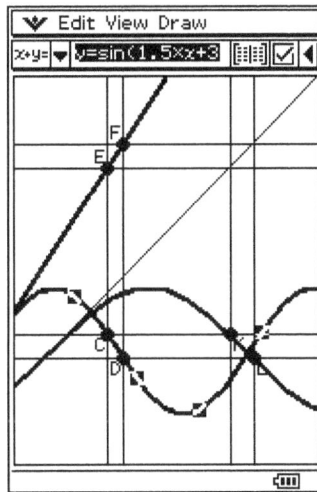

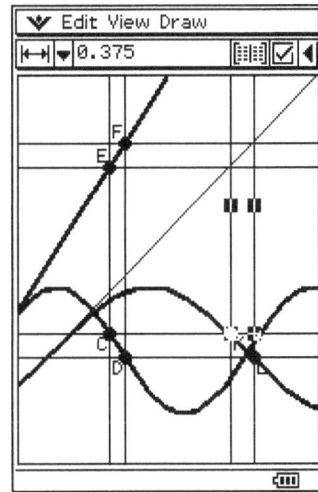

As K and L are the same height as C and D, the ratio of the areas of the trapezoids under these points is equal to the ratio of the horizontal distances. You can confirm this by creating the trapezoids:

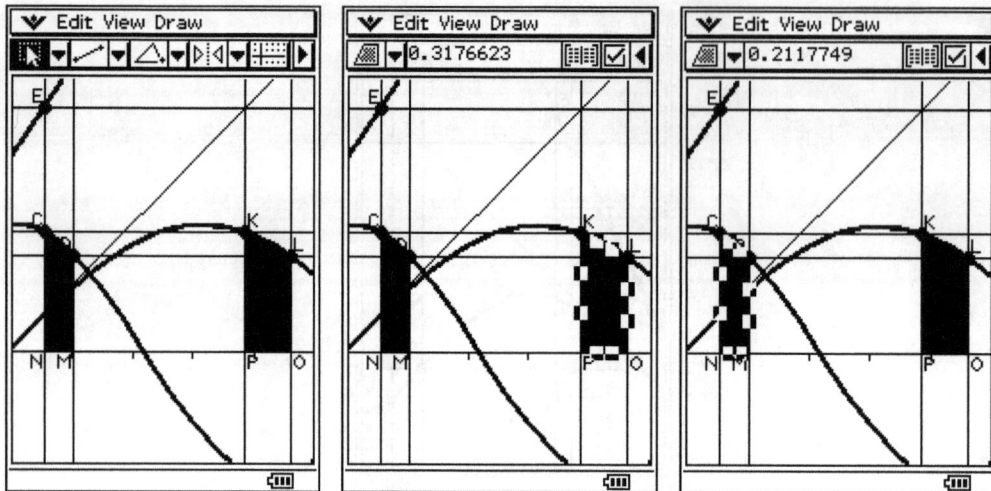

We see that the ratio of the areas is indeed the slope of AB:

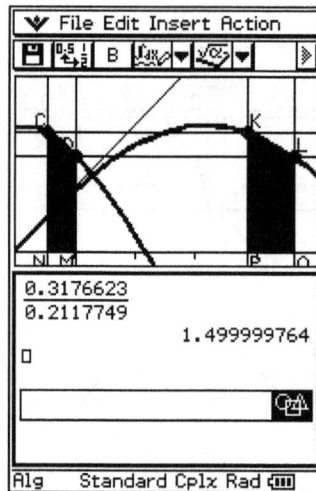

Hence we have the following:

$$Area(CD) = \frac{Area(KL)}{Slope(AB)}$$

## Some Trig Integrals

We can apply the formula:

$$\int f(a.x+b)dx = \frac{1}{a}\int f(x)dx$$

directly to compute integrals such as the following:

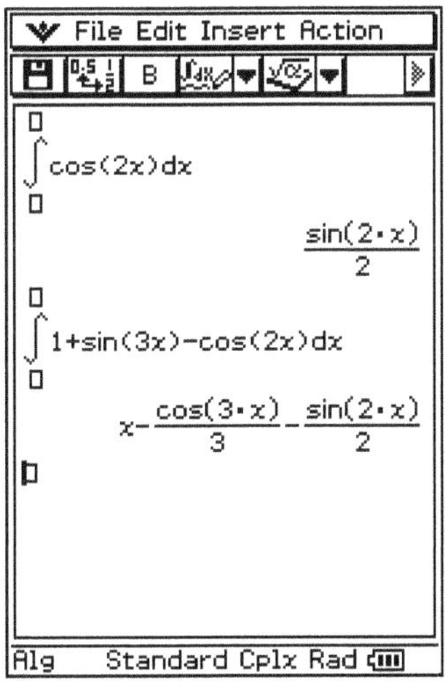

A little trigonometric pre-processing (which ClassPad does embarrassingly easily) can, however, extend the range of this approach to products of trig functions:

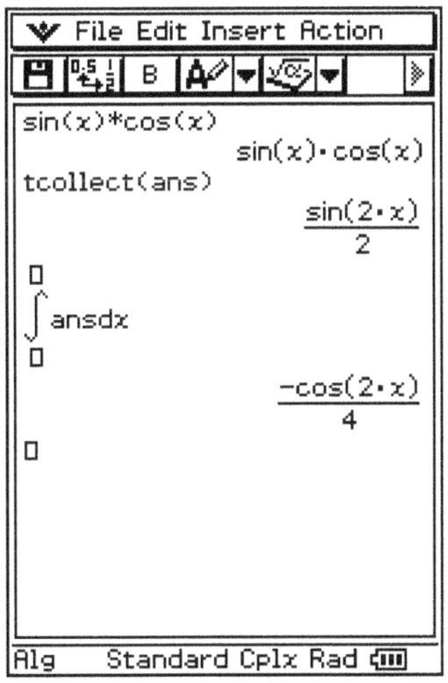

We can apply the same technique to relatively complicated trig products.

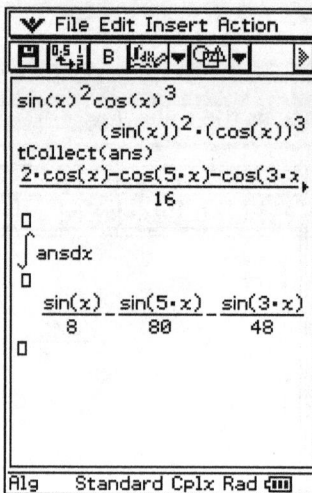

```
  ▼ File Edit Insert Action
 ┌──┬───┬─┬─────┬─────┬────┬──┐
 │🖫 │0.5│B│📊▼ │📊▼ │    │ ▷│
 └──┴───┴─┴─────┴─────┴────┴──┘
 sin(x)²cos(x)³
           (sin(x))²·(cos(x))³
 tCollect(ans)
   2·cos(x)-cos(5·x)-cos(3·x ▸
 ─────────────────────────────
             16
  □
  ⌠
  │ ansdx
  ⌡
  □
    sin(x)   sin(5·x)   sin(3·x)
    ────── - ──────── - ────────
      8         80        48
  □

 ┌─────────────────────────────┐
 Alg    Standard Cplx Rad 🔋
```

EXERCISES

Evaluate the following integrals.  Check your answers in ClassPad:

1. $\displaystyle\int \cos(2x)\,dx$

2. $\displaystyle\int e^{3x-2}\,dx$

3. $\displaystyle\int \sqrt{\frac{x}{2}-1}\,dx$

4. $\displaystyle\int \cos(x)^3\,dx$

5. $\displaystyle\int \sin(x)^2 \cos(x)^2\,dx$

## Integration by Parts

The product rule for differentiation says the following:

$$\frac{d}{dx}\big(u(x) \cdot v(x)\big) = \frac{du}{dx} \cdot v + u\frac{dv}{dx}$$

If we integrate this equation (both sides), we get the following:

$$u \cdot v = \int \frac{du}{dx} \cdot v dx + \int u\frac{dv}{dx}dx$$

Which can be rewritten:

$$\int u\frac{dv}{dx}dx = u \cdot v - \int \frac{du}{dx} \cdot v dx$$

If we are asked to integrate a function f(x).g(x), and we know the integral of g(x), then we set:

$$u = f$$

$$v = \int g dx$$

And apply the above formula as follows:

$$\int f \cdot g dx = f \cdot \int g dx - \int \left( \frac{df}{dx} \cdot \int g dx \right) dx$$

Notice, we still have an integral to deal with, and in general this new integral will be more complicated than the original, so we will not have made any progress. However, there are some families of functions for which this method works like a charm. Let's examine some examples:

$$\int x \cdot \sin(x)$$

$$\int x^2 \cdot \cos(x)$$

---

$$\int x \cdot e^x$$

$$\int \sin(x) \cdot e^x$$

We'll look at the first of these. First we define f(x) and g(x), then apply the integration by parts formula:

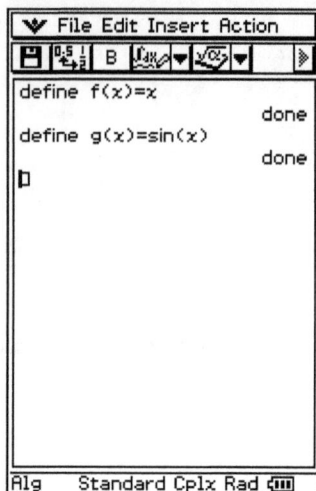

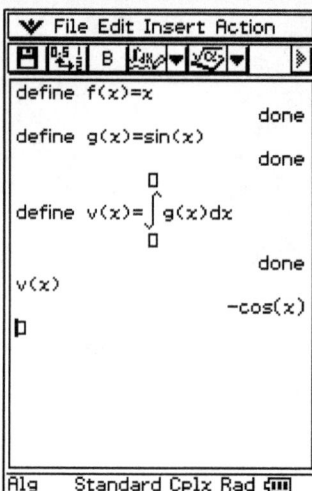

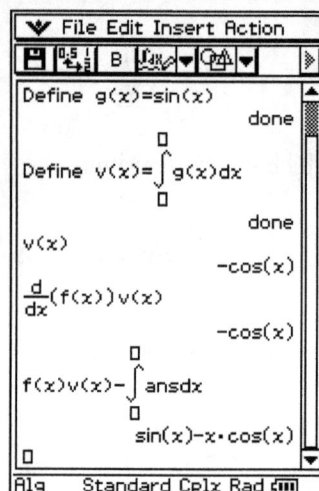

| | | |
|---|---|---|
| ▼ File Edit Insert Action ≫ | ▼ File Edit Insert Action ≫ | ▼ File Edit Insert Action ≫ |
| define f(x)=x<br>　　　　　done<br>define g(x)=sin(x)<br>　　　　　done<br>▯ | define f(x)=x<br>　　　　　done<br>define g(x)=sin(x)<br>　　　　　done<br>define v(x)=∫g(x)dx<br>　　　　　done<br>v(x)<br>　　　　　−cos(x)<br>▯ | Define g(x)=sin(x)<br>　　　　　done<br>Define v(x)=∫g(x)dx<br>　　　　　done<br>v(x)<br>　　　　　−cos(x)<br>$\frac{d}{dx}$(f(x))v(x)<br>　　　　　−cos(x)<br>f(x)v(x)−∫ansdx<br>　　sin(x)−x·cos(x)<br>▯ |
| Alg　Standard Cplx Rad ▥ | Alg　Standard Cplx Rad ▥ | Alg　Standard Cplx Rad ▥ |

As usual, we can check our answer by differentiation:

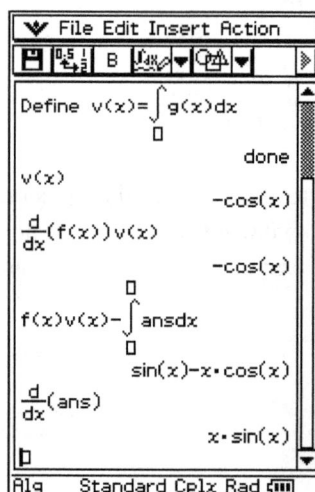

| |
|---|
| ▼ File Edit Insert Action ≫ |
| Define v(x)=∫g(x)dx<br>　　　　　done<br>v(x)<br>　　　　　−cos(x)<br>$\frac{d}{dx}$(f(x))v(x)<br>　　　　　−cos(x)<br>f(x)v(x)−∫ansdx<br>　　sin(x)−x·cos(x)<br>$\frac{d}{dx}$(ans)<br>　　　　　x·sin(x)<br>▯ |
| Alg　Standard Cplx Rad ▥ |

We can do the second example simply by changing the definitions of f(x) and g(x).

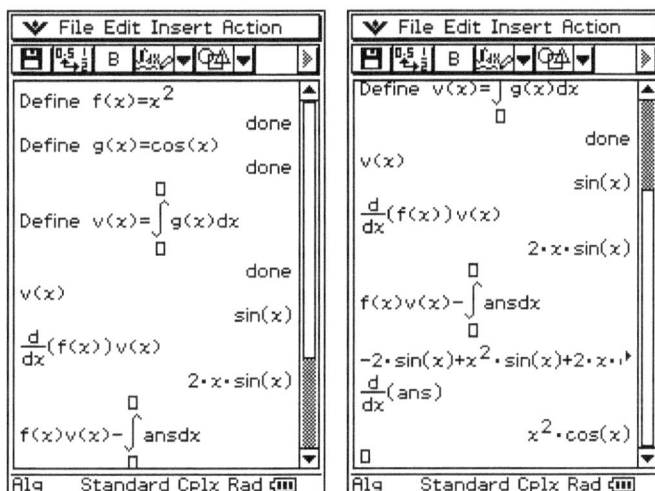

Notice that the integration by parts formula requires us to solve:

$$\int x \cdot \sin(x)dx$$

which we integrated in the previous example.

Let's look at a general problem of this type:

$$\int x^n \cdot \sin(x)dx$$

Observe that the integration by parts formula reduces the order of n (as well as changing the sin to a cos).

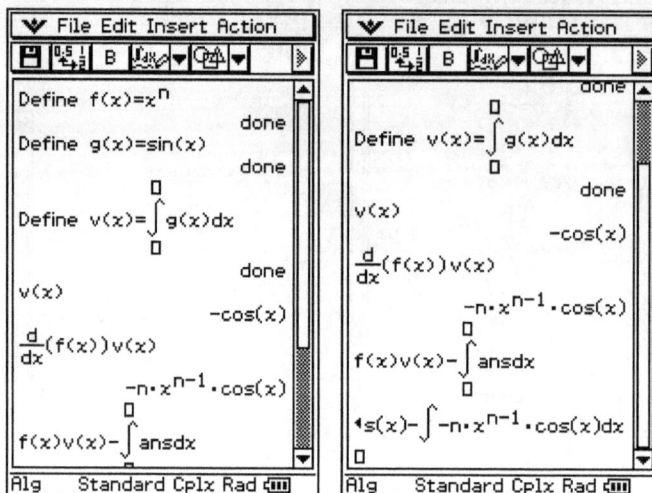

Applying the method n times will eventually reduce n to 0 and finally let us integrate $\cos(x)$ (or $\sin(x)$ depending on whether n is odd or even).

While we have this eActivity on the screen, we may as well dispose of the similar example involving $e^x$.

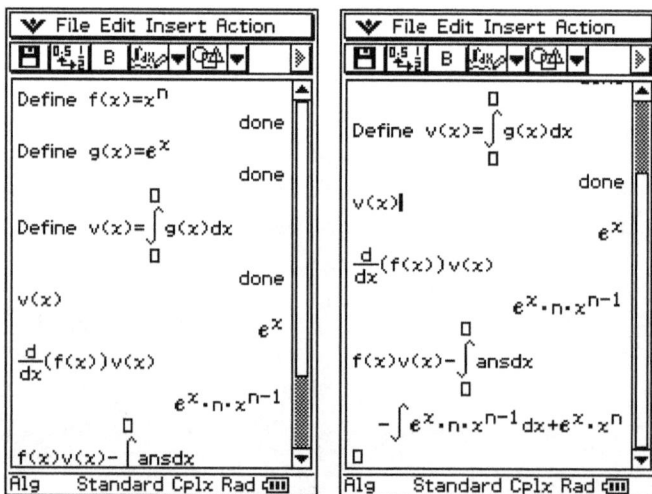

Apply the result again:

File Edit Insert Action

$$e^x$$

$$\frac{d}{dx}(f(x))v(x)$$

$$e^x \cdot n \cdot x^{n-1}$$

$$f(x)v(x)-\int ansdx$$

$$-\int e^x \cdot n \cdot x^{n-1}dx+e^x \cdot x^n$$

$$e^x \cdot x^n - e^x \cdot n \cdot x^{n-1}+\int e^x \cdot \blacktriangleright$$

$$\int e^x \cdot n \cdot (n-1) \cdot x^{n-2}dx+e^x \blacktriangleright$$

Alg    Standard Cplx Rad

Can you come up with the general formula (after the integral has finally been worked out of the expression)? You can confirm you are correct by evaluating it for one or two values of n, and differentiating the result:

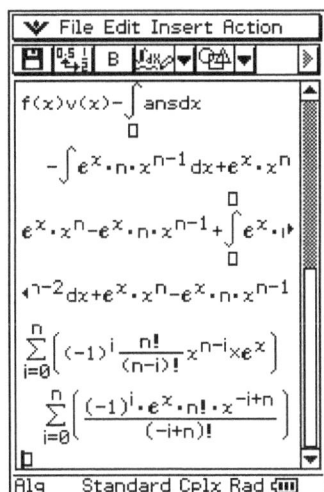

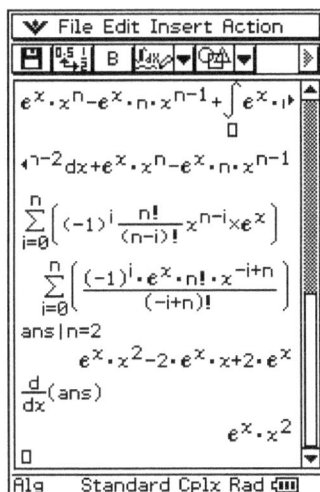

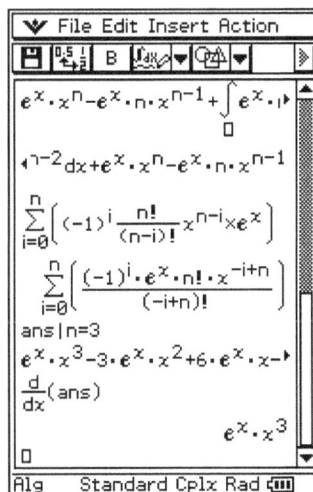

File Edit Insert Action

$$f(x)v(x)-\int ansdx$$

$$-\int e^x \cdot n \cdot x^{n-1}dx+e^x \cdot x^n$$

$$e^x \cdot x^n - e^x \cdot n \cdot x^{n-1}+\int e^x \cdot \blacktriangleright$$

$$\cdots^{n-2}dx+e^x \cdot x^n - e^x \cdot n \cdot x^{n-1}$$

$$\sum_{i=0}^{n}\left((-1)^i \frac{n!}{(n-i)!}x^{n-i}\times e^x\right)$$

$$\sum_{i=0}^{n}\left(\frac{(-1)^i \cdot e^x \cdot n! \cdot x^{-i+n}}{(-i+n)!}\right)$$

Alg    Standard Cplx Rad

File Edit Insert Action

$$e^x \cdot x^n - e^x \cdot n \cdot x^{n-1}+\int e^x \cdot \blacktriangleright$$

$$\cdots^{n-2}dx+e^x \cdot x^n - e^x \cdot n \cdot x^{n-1}$$

$$\sum_{i=0}^{n}\left((-1)^i \frac{n!}{(n-i)!}x^{n-i}\times e^x\right)$$

$$\sum_{i=0}^{n}\left(\frac{(-1)^i \cdot e^x \cdot n! \cdot x^{-i+n}}{(-i+n)!}\right)$$

ans|n=2

$$e^x \cdot x^2 - 2 \cdot e^x \cdot x + 2 \cdot e^x$$

$$\frac{d}{dx}(ans)$$

$$e^x \cdot x^2$$

Alg    Standard Cplx Rad

File Edit Insert Action

$$e^x \cdot x^n - e^x \cdot n \cdot x^{n-1}+\int e^x \cdot \blacktriangleright$$

$$\cdots^{n-2}dx+e^x \cdot x^n - e^x \cdot n \cdot x^{n-1}$$

$$\sum_{i=0}^{n}\left((-1)^i \frac{n!}{(n-i)!}x^{n-i}\times e^x\right)$$

$$\sum_{i=0}^{n}\left(\frac{(-1)^i \cdot e^x \cdot n! \cdot x^{-i+n}}{(-i+n)!}\right)$$

ans|n=3

$$e^x \cdot x^3 - 3 \cdot e^x \cdot x^2 + 6 \cdot e^x \cdot x - \blacktriangleright$$

$$\frac{d}{dx}(ans)$$

$$e^x \cdot x^3$$

Alg    Standard Cplx Rad

Our last integral is this:

$$\int \sin(x) \cdot e^x$$

```
▼ File Edit Insert Action
[icons]                        ▷
Define f(x)=sin(x)
                         done
Define g(x)=eˣ
                         done
             □
Define v(x)=∫ g(x)dx
             □
                         done
v(x)
                       eˣ
d
──(f(x))v(x)
dx
                   eˣ·cos(x)
□
Alg    Standard Cplx Rad ▦
```

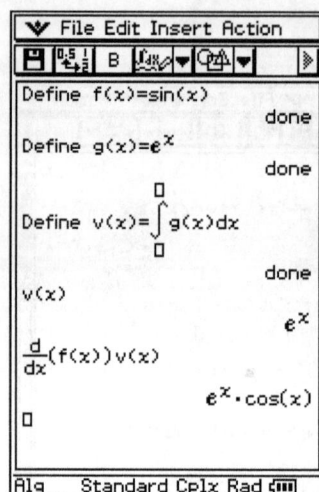

Applying the integration by parts formula converts this to an integral of the form:

$$\int \cos(x)\cdot e^x$$

Clearly, if we apply it again we'll get back to where we started:

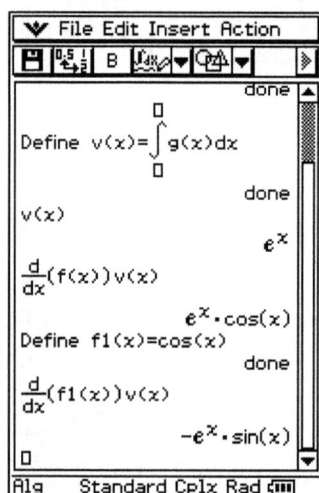

```
▼ File Edit Insert Action
[icons]                        ▷
                     done ▲
          □
Define v(x)=∫ g(x)dx
          □
                    done
v(x)
                   eˣ
d
──(f(x))v(x)
dx
                eˣ·cos(x)
Define f1(x)=cos(x)
                   done
d
──(f1(x))v(x)
dx
              -eˣ·sin(x)
□                       ▼
Alg    Standard Cplx Rad ▦
```

However, if we let I be the desired integral, we can use the integration by parts formula to give us an equation for I:

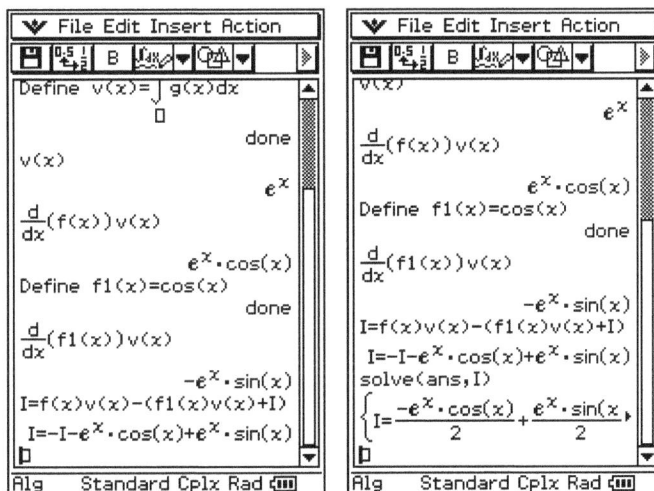

We can check this equation by letting ClassPad do the original integral:

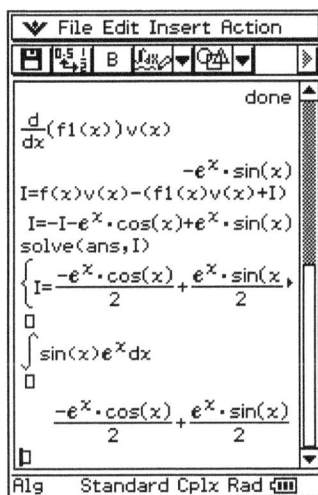

EXERCISES

Use integration by parts to evaluate the following integrals. Check your answers using ClassPad:

1. $\int x \cos(2x)\,dx$

2. $\int x^2 \sin(x)\,dx$

3. $\int x e^{kx}\,dx$

4. $\int \sin x e^x\,dx$

# Change of Variable

In a previous section, we looked at

$$\int f(a \cdot x + b)dx$$

In this section we'd like to examine the more general problem:

$$\int f(g(x))dx$$

Again let's look at a specific example and examine the situation geometrically.

We'll study the function:

$$y(x) = \sin(\frac{x^2}{2} + \frac{3}{2})$$

Draw this function, along with the component functions:

$$g(x) = \frac{x^2}{2} + \frac{3}{2}$$

$$f(x) = \sin(x)$$

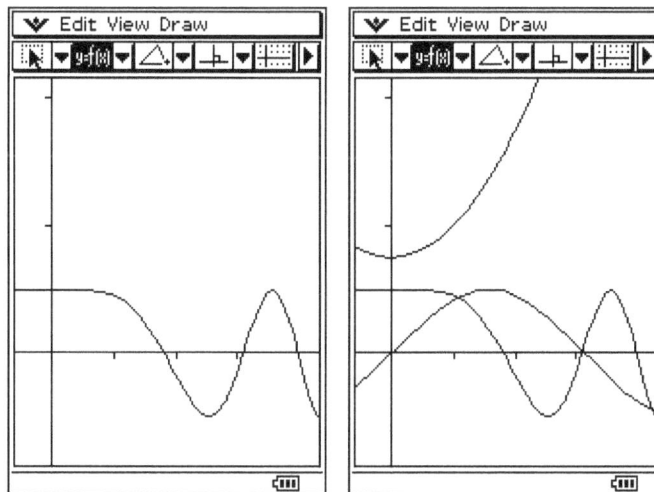

Ideally, we would at this point create two points on the function y(x), draw vertical lines through those points and intersect them with the function g(x). Unfortunately, while ClassPad will let you intersect two lines, it will not let you intersect a line and a function. (The reason for this is that in general – for non vertical lines – intersecting a line and a function is difficult.)

With a little cleverness, however, we can still create our model. What we can do is to draw two points A and B on y(x) and two points C and D on g(x), and synchronize their location by animating them between the same limits.

Start off with A and B, add animations and use **Edit/Animation/Edit Animations** to specify limits. (Making the points thicker helps in understanding the drawing.)

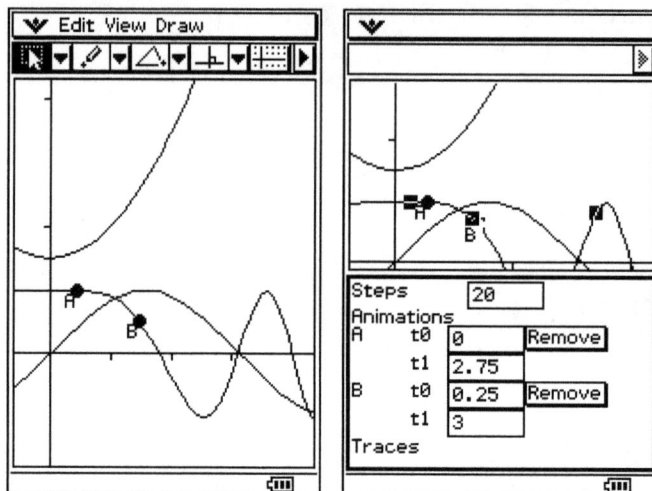

Draw points C and D on the curve g(x), add animations and specify the same limits for these points as for A and B:

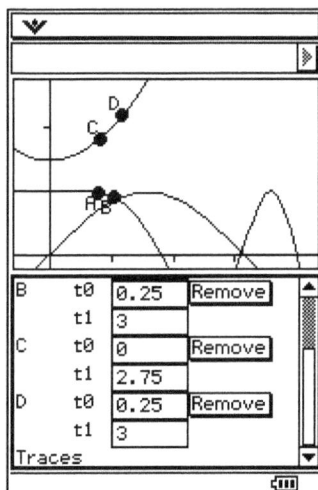

Notice, they do not automatically line up. But they will when you run an animation. The animation will force A and C to walk along their respective curves at the same x location. Likewise B and D:

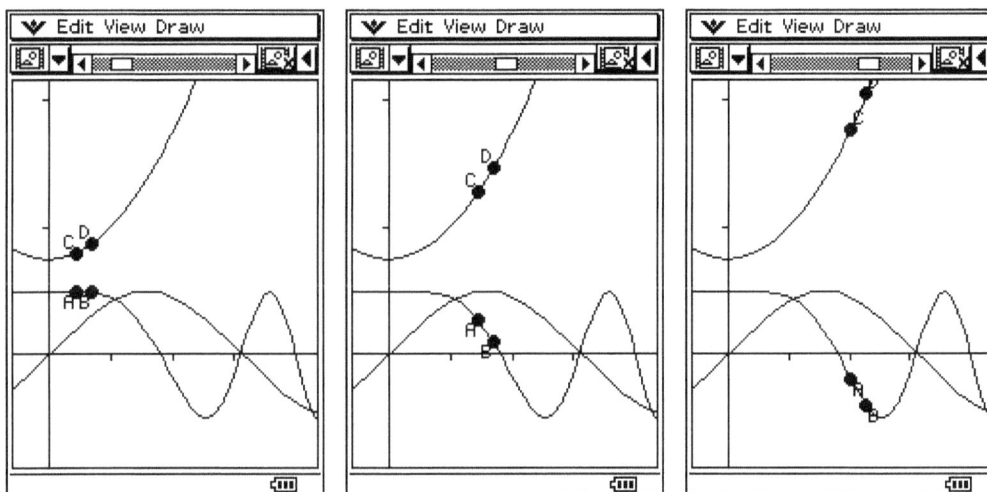

Let's put lines through AC and BD, first thickening the existing curves to help us keep seeing them:

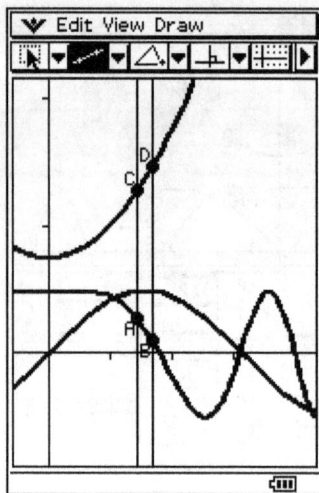

Next, we draw a line EF and constrain it to have the equation y=x:

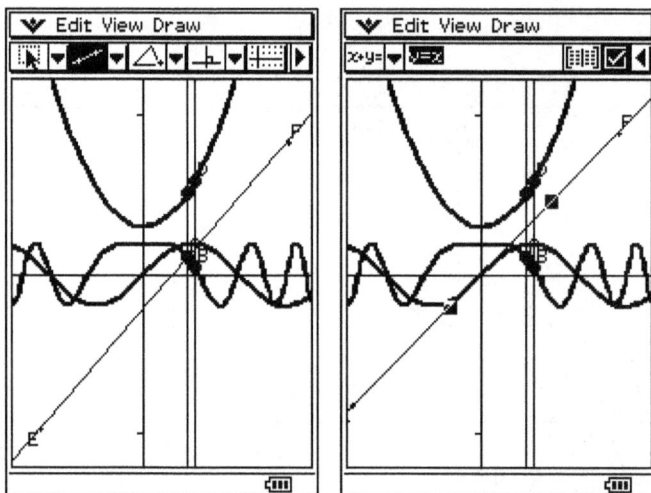

Now we draw lines through C and D and constrain them to be horizontal:

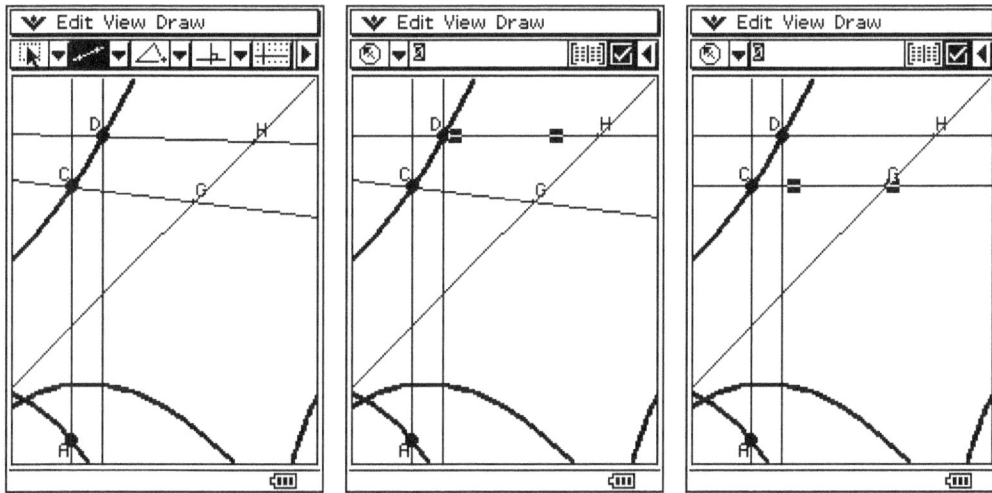

Reflect these lines in y=x and intersect them with horizontals constructed from A and B:

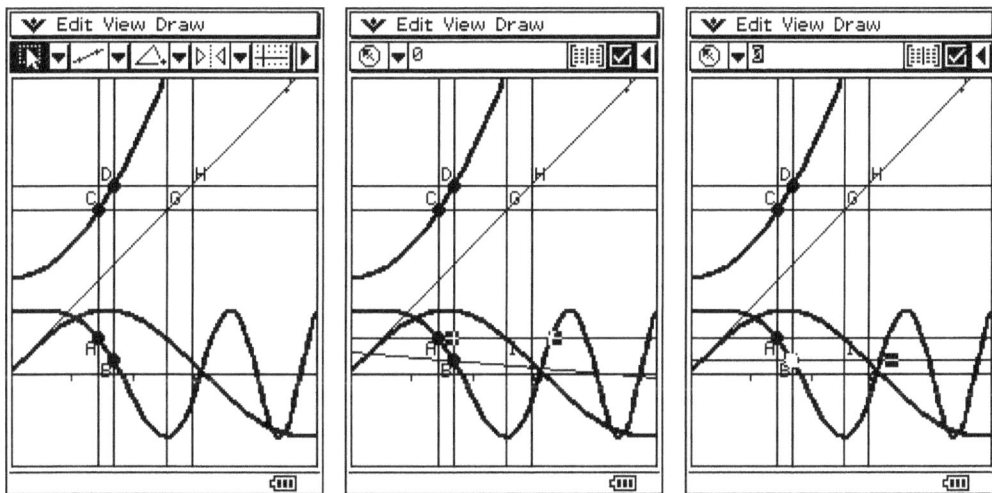

Make the new points I and J thicker so you can see them. Now run the animation and observe that as A and B traverse the function y(x), I and J traverse the function f(x):

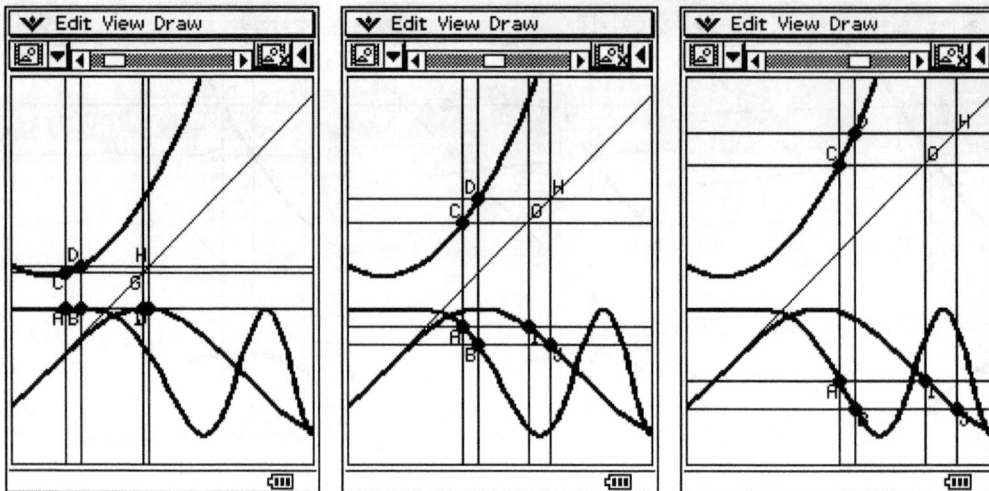

Notice that the spacing of I and J increases as points C and D get to steeper parts of the quadratic.

Let r be the difference in x coordinates between A and B, and let s be the difference in x coordinates between I and J. As A is the same height as I and B is the same height as J, the ratio of the area under IJ to the area under AB is:

$$\frac{Area(IJ)}{Area(AB)} = \frac{s}{r}$$

We can see from the diagram that s is equal to the difference in y coordinates between C and D. Hence

$$\frac{s}{r} = Slope(CD)$$

Hence the ratio of the areas is simply the slope of the chord CD. You can verify this from the diagram:

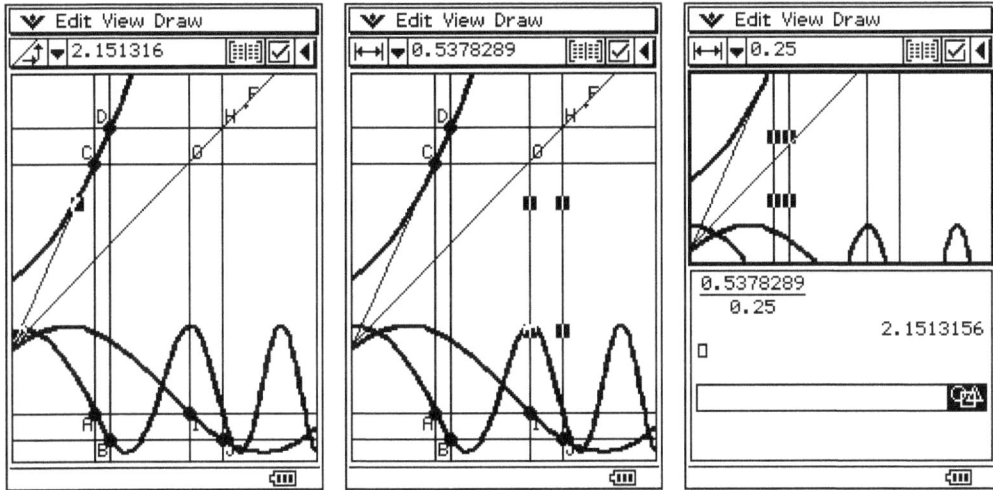

Hence, the area AB can be expressed as follows:

$$Area(AB) = \frac{r}{s} Area(IJ) = \frac{Area(IJ)}{Slope(CD)}$$

Moving from trapezoids and chords to integrals and tangents, the analogous formula is:

$$\int f(g(x))dx = \int \left(\frac{dg}{dx}\right)^{-1} f(g)dg$$

This can be derived from the chain rule as follows:

Let

$$u(x) = \int f(x)dx$$

The fundamental theorem of calculus tells us that:

$$\frac{du}{dx} = f$$

```
▼ File Edit Insert Action
[icons]                          »

              □
Define u(x)=∫ f(x)dx
              □
                            done
 d
───(u(x))
 dx
                            f(x)
 □


Alg    Standard Cplx Rad (▯▯)
```

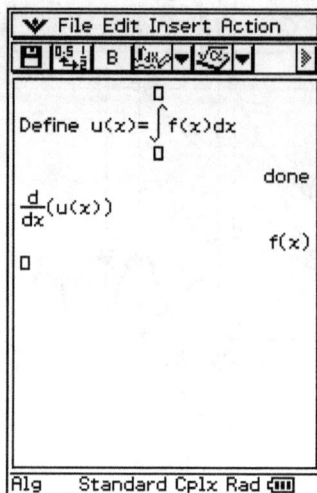

The chain rule says:

$$\frac{du}{dx} = \frac{du}{dg}\frac{dg}{dx}$$

Hence

$$\frac{du}{dg} = \left(\frac{dg}{dx}\right)^{-1}\frac{du}{dx}$$

$$\Rightarrow \int\frac{du}{dg}dg = \int\left(\frac{dg}{dx}\right)^{-1}\frac{du}{dx}dg$$

$$\Rightarrow u = \int\left(\frac{dg}{dx}\right)^{-1}fdg$$

$$\Rightarrow \int f(x)dx = \int\left(\frac{dg}{dx}\right)^{-1}fdg$$

## Applying Change of Variable

The above change of variable formula is all very well, but there is still an integral to be done. Not only that, the thing to be integrated has just become (typically) more complicated. So the change of variable technique only really helps in situations where some cancellation occurs (as frequently happens in contrived examples).

Let's contrive some examples. Apply the change of variables technique to find these integrals:

$$\int x^2 e^{x^3} dx$$

$$\int \sin(x)\cos^2(x) dx$$

$$\int \frac{\sin(x)}{\cos(x)} dx$$

We'll define f(x) to be the function whose integral we are seeking, and g(x) to be the new variable, then enter the change of variable formula:

The next step is to integrate with respect to g. However, if you try this directly, you'll get an error message:

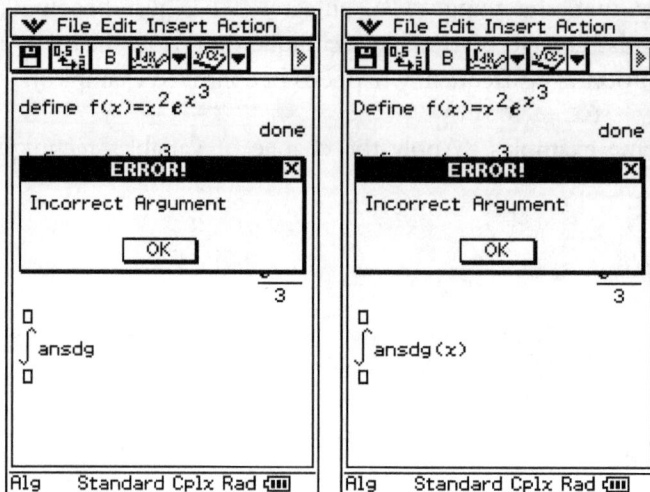

```
▼ File Edit Insert Action          ▶
💾 📊 B ⌨▼ √▦▼              »

define f(x)=x²eˣ³
                          done

       ┌─ ERROR! ──────── ✕ ┐
       │ Incorrect Argument  │
       │      ┌────────┐     │
       │      │   OK   │     │
       └──────┴────────┴─────┘
                          3
  ☐
  ∫ ansdg
  ☐

Alg    Standard Cplx Rad ▭
```

```
▼ File Edit Insert Action          ▶
💾 📊 B ⌨▼ √▦▼              »

Define f(x)=x²eˣ³
                          done

       ┌─ ERROR! ──────── ✕ ┐
       │ Incorrect Argument  │
       │      ┌────────┐     │
       │      │   OK   │     │
       └──────┴────────┴─────┘
                          3
  ☐
  ∫ ansdg(x)
  ☐

Alg    Standard Cplx Rad ▭
```

ClassPad will not allow you to integrate with respect to a function, it wants you to integrate with respect to a variable. You can do this by substitution – replace g(x) by a dummy variable u, integrate with respect to u, then replace u with g(x):

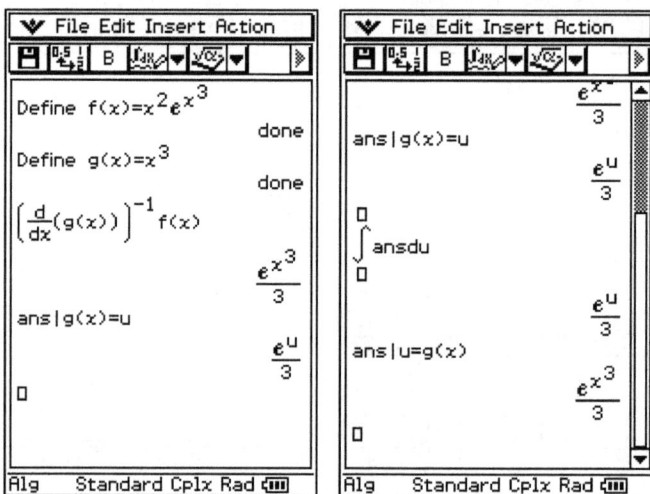

```
▼ File Edit Insert Action          ▶
💾 📊 B ⌨▼ √▦▼              »

Define f(x)=x²eˣ³
                          done
Define g(x)=x³
                          done
  ⎛ d        ⎞⁻¹
  ⎜ ──(g(x)) ⎟   f(x)
  ⎝ dx       ⎠
                        eˣ³
                        ───
                         3
ans|g(x)=u
                        eᵘ
                        ──
                         3
  ☐

Alg    Standard Cplx Rad ▭
```

```
▼ File Edit Insert Action          ▶
💾 📊 B ⌨▼ √▦▼              »

                        eˣ˙ ▲
                        ───
                         3
ans|g(x)=u

                        eᵘ
                        ──
  ☐                      3
  ∫ ansdu
  ☐
                        eᵘ
                        ──
                         3
ans|u=g(x)
                        eˣ³
                        ───
  ☐                      3
                          ▼
Alg    Standard Cplx Rad ▭
```

We can check our answer by differentiating it:

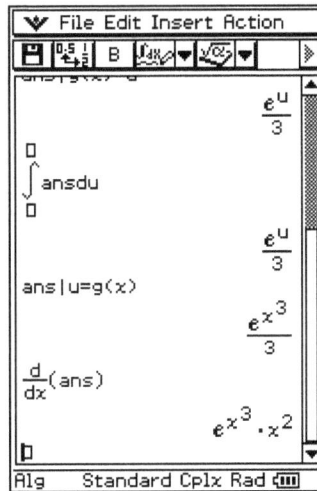

Let's look at the second example. We set new values for f and g, and we see that the rest of the page updates appropriately:

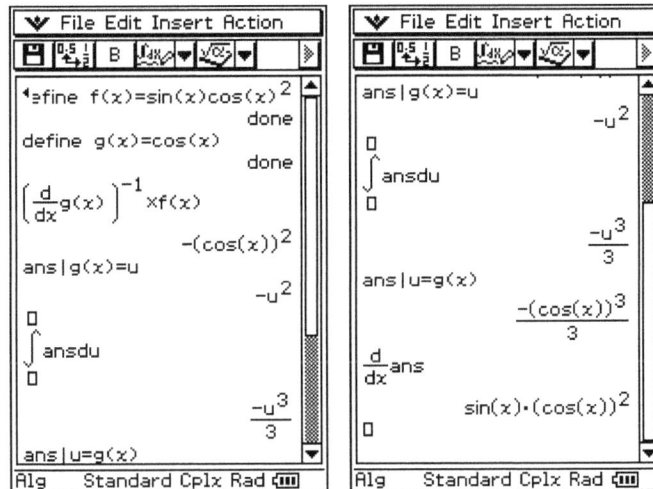

Do the same for the final example. Notice that this is a way to find the integral of tan(x).

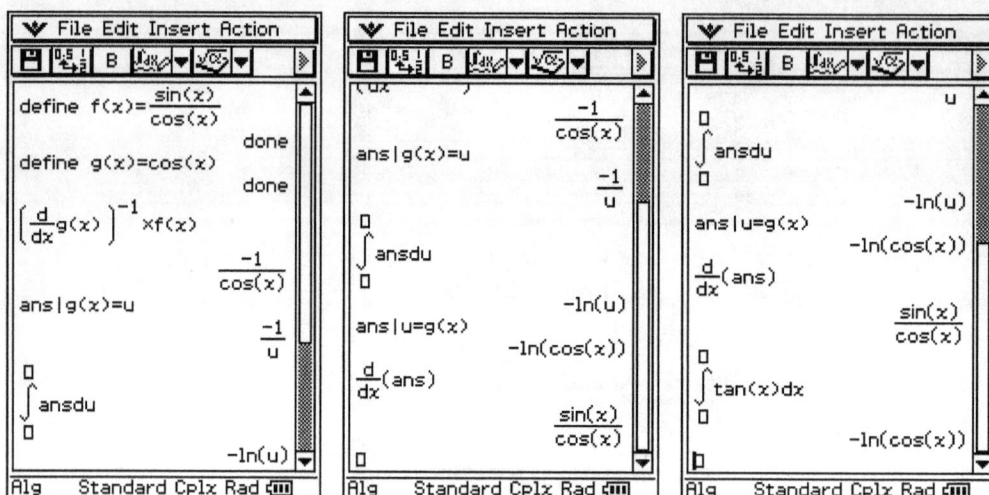

The above is an example of a more general form, where:

$$f(x) = \frac{1}{h(x)} \frac{dh}{dx}$$

We can apply the same analysis, just by changing the definitions of f(x) and g(x) appropriately:

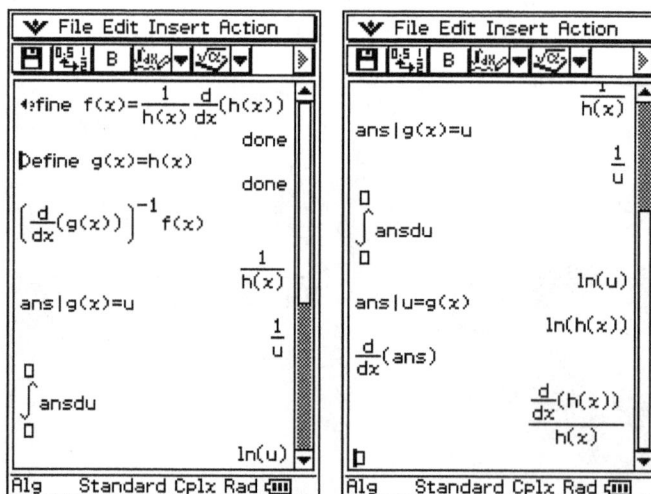

In all of the above examples, the x terms luckily disappeared from the expression before we had to integrate with respect to u. If x persists, we cannot ignore it, we need to substitute for x as a function of u before integrating.

Let's look at an example:

$$\int \sqrt{x} e^{\sqrt{x}} dx$$

Following our previous method, we enter f(x) and g(x) and the variable substitution expression:

Unlike previous examples, however, we notice that the x has not disappeared from our expression. It has to before we can integrate with respect to u. So we need to substitute for x:

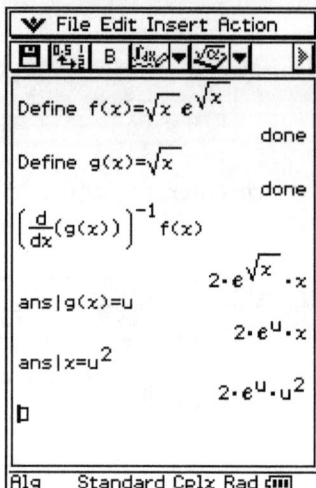

```
▼ File Edit Insert Action
[icons]

Define f(x)=√x e^√x
                                done
Define g(x)=√x
                                done
(d/dx(g(x)))^-1 f(x)
                        2·e^√x ·x
ans|g(x)=u
                        2·e^u·x
ans|x=u^2
                        2·e^u·u^2
▯

Alg     Standard Cplx Rad
```

This is an expression which we can integrate by parts (or just use ClassPad's integrate function). We then need to substitute back in for u in terms of x:

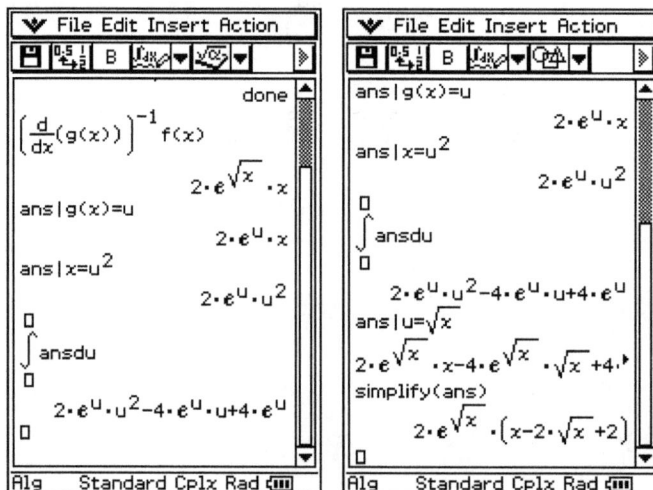

```
▼ File Edit Insert Action
[icons]
                        done
(d/dx(g(x)))^-1 f(x)
                        2·e^√x ·x
ans|g(x)=u
                        2·e^u·x
ans|x=u^2
                        2·e^u·u^2
▯
∫ ansdu
▯
        2·e^u·u^2-4·e^u·u+4·e^u
▯

Alg     Standard Cplx Rad
```

```
▼ File Edit Insert Action
[icons]
ans|g(x)=u
                        2·e^u·x
ans|x=u^2
                        2·e^u·u^2
▯
∫ ansdu
▯
        2·e^u·u^2-4·e^u·u+4·e^u
ans|u=√x
2·e^√x ·x-4·e^√x ·√x +4·▸
simplify(ans)
                2·e^√x ·(x-2·√x +2)
▯

Alg     Standard Cplx Rad
```

The answer can be checked by differentiation:

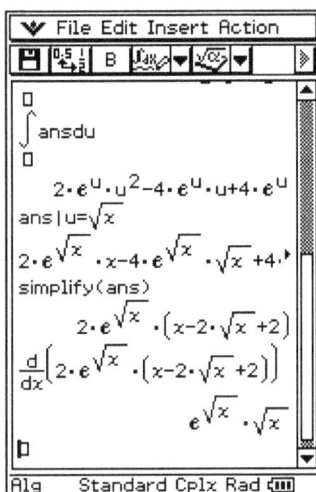

EXERCISES

Integrate the following. Check by differentiating your answer in ClassPad.

1. $\int x \sin(x^2)\,dx$

2. $\int \dfrac{\cos(\sqrt{x})}{\sqrt{x}}\,dx$

3. $\int \dfrac{x}{\sqrt{1-x^2}}\,dx$

# Introducing Your Own Variables

The above examples relied on the fact that the integrand had a derivative already hidden in it. Integration gets really creative when you dream up a function to substitute for x.

In such cases, we want to substitute for x as some function of t:

$$x = x(t)$$

Applying the variable substitution formula:

$$\int f dx = \int \left(\frac{dt}{dx}\right)^{-1} f(t) dt$$

From the chain rule, we know the following:

$$1 = \frac{dx}{dx} = \frac{dx}{dt}\frac{dt}{dx}$$

Hence:

$$\frac{dx}{dt} = \left(\frac{dt}{dx}\right)^{-1}$$

And the variable substitution formula can be written:

$$\int f dx = \int \frac{dx}{dt} f(t) dt$$

Which is a more convenient form if you are introducing your own variables.

Let's try the following integral:

$$\int \sqrt{4 - x^2}\, dx$$

The clever substitution we can try for x is x=2sin(t). With this substitution, the integrand becomes 2cos(t) and there is some hope of suitable simplification occurring.

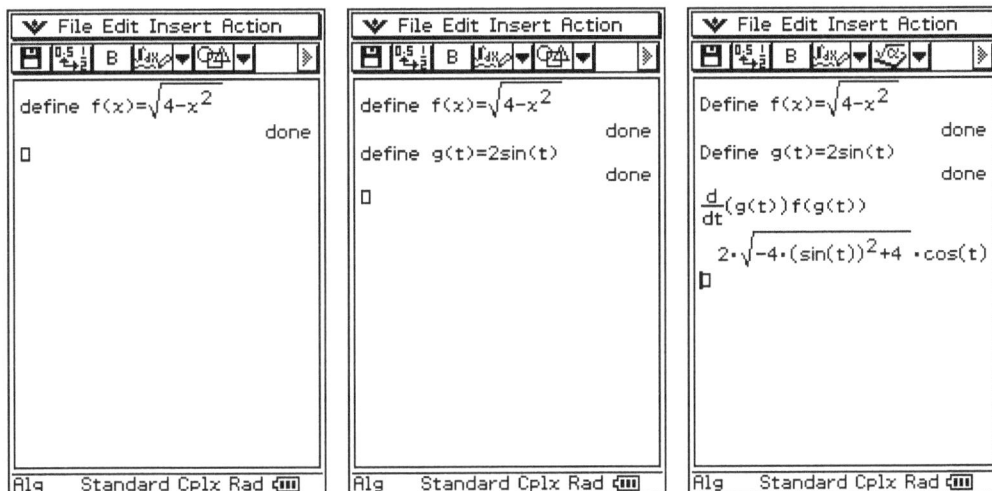

ClassPad is a little reluctant to recognize that the root term is just 2cos(t). We can help it make the simplification by evaluating the answer to the previous computation with the root set to be equal to 2cos(t):

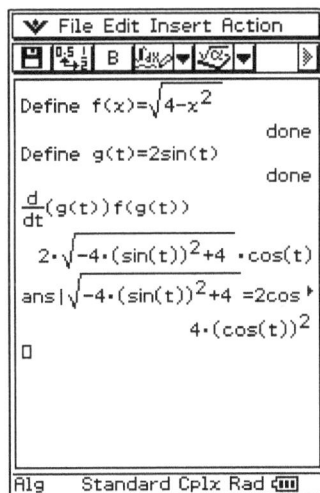

Now tcollect() will make the integration look easier:

| ▼ File Edit Insert Action | ▼ File Edit Insert Action |
|---|---|
| Define f(x)=$\sqrt{4-x^2}$ | done |
| done | Define g(t)=2sin(t) |
| Define g(t)=2sin(t) | done |
| done | $\frac{d}{dt}$(g(t))f(g(t)) |
| $\frac{d}{dt}$(g(t))f(g(t)) | $2\cdot\sqrt{-4\cdot(\sin(t))^2+4}\cdot\cos(t)$ |
| $2\cdot\sqrt{-4\cdot(\sin(t))^2+4}\cdot\cos(t)$ | ans$\|\sqrt{-4\cdot(\sin(t))^2+4}=2\cos$ |
| ans$\|\sqrt{-4\cdot(\sin(t))^2+4}=2\cos$ | $4\cdot(\cos(t))^2$ |
| $4\cdot(\cos(t))^2$ | tCollect(ans) |
| tCollect(ans) | $2\cdot\cos(2\cdot t)+2$ |
| $2\cdot\cos(2\cdot t)+2$ | □ |
| □ | $\int$ ansdt |
| | □ |
| | $2\cdot t+\sin(2\cdot t)$ |
| | □ |
| Alg    Standard Cplx Rad | Alg    Standard Cplx Rad |

Afer integrating, we need to substitute back in for t.  texpand() can be used to make the answer look a little more attractive:

| ▼ File Edit Insert Action | ▼ File Edit Insert Action |
|---|---|
| dt | $4\cdot(\cos(t))^2$ |
| $2\cdot\sqrt{-4\cdot(\sin(t))^2+4}\cdot\cos(t$ | tcollect(ans) |
| ans$\|\sqrt{-4\cdot(\sin(t))^2+4}=2\cos$ | $2\cdot\cos(2\cdot t)+2$ |
| $4\cdot(\cos(t))^2$ | □ |
| tcollect(ans) | $\int$ ansdt |
| $2\cdot\cos(2\cdot t)+2$ | □ |
| □ | $2\cdot t+\sin(2\cdot t)$ |
| $\int$ ansdt | ans$\|t=\sin^{-1}(\frac{x}{2})$ |
| □ | $2\cdot\sin^{-1}\left(\frac{x}{2}\right)+\sin\left(2\cdot\sin^{-1}\left(\frac{x}{2}\right)\right)$ |
| $2\cdot t+\sin(2\cdot t)$ | texpand(ans) |
| ans$\|t=\sin^{-1}(\frac{x}{2})$ | $2\cdot\sin^{-1}\left(\frac{x}{2}\right)+\sqrt{\frac{-x^2}{4}+1}\cdot x$ |
| $2\cdot\sin^{-1}\left(\frac{x}{2}\right)+\sin\left(2\cdot\sin^{-1}\left(\frac{x}{2}\right)\right)$ | |
| □ | □ |
| Alg    Standard Cplx Rad | Alg    Standard Cplx Rad |

We can verify our result by differentiation, or just by letting ClassPad integrate the original function:

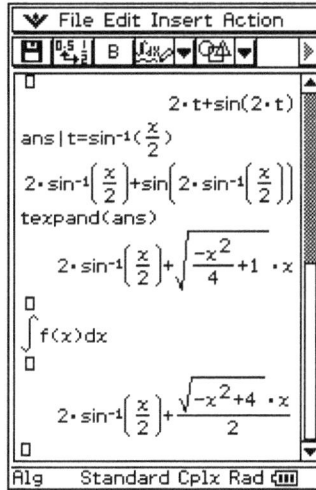

EXERCISES

1. Find $\int \sqrt{1 - x^2}\, dx$

2. Find $\int \sqrt{\dfrac{x^2}{4} - 1}\, dx$

3. What is the derivative of $\sinh^{-1}(x)$

4. Find $\int \sqrt{1 + x^2}\, dx$

## Integrating Rational Functions

There are a suite of techniques which we can apply to evaluate integrals of the form:

$$\int \frac{P(x)}{Q(x)} dx$$

where P(x) and Q(x) are polynomials.

Here are some examples:

$$\int \frac{1}{3x-2} dx$$

$$\int \frac{2x+3}{x-1} dx$$

$$\int \frac{2x-3}{x^2-3x-1} dx$$

$$\int \frac{x+1}{x^2-x-2} dx$$

The first is simple enough that we can let ClassPad do it without endangering our comprehension:

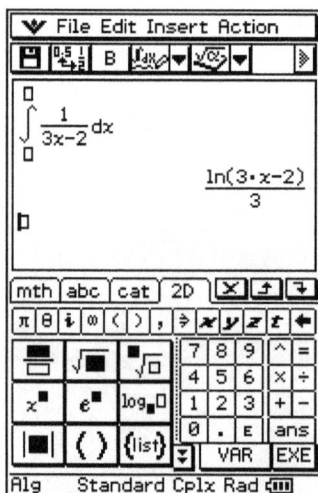

The second example requires a small amount of preparation. Both the numerator and denominator are linear functions. We can perform polynomial division to convert this to a constant plus a fraction whose numerator is constant. The algebra function propFrac() does this for us. The resulting function can readily be integrated:

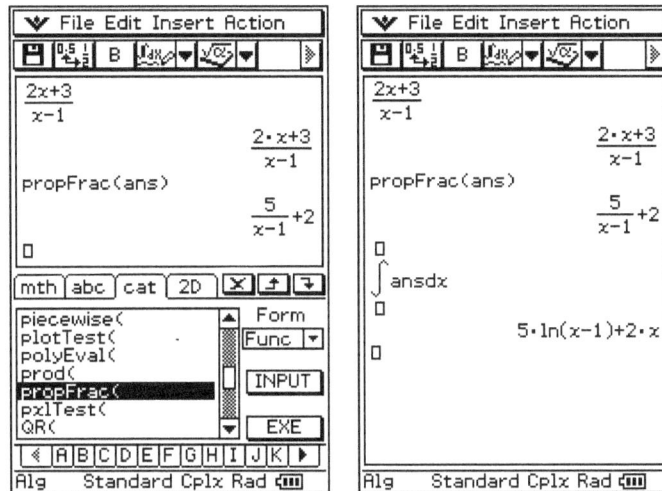

In the next example, we are lucky – the top line is just the derivative of the bottom line. So we can use the method of variable substitution:

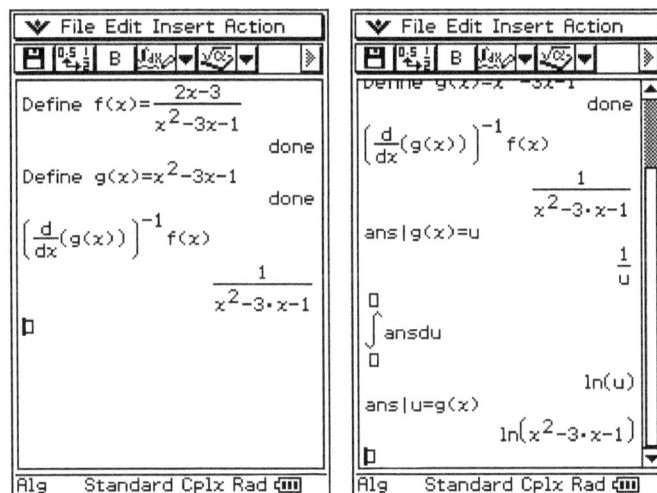

In the next example, we do not have such good luck. Attempting to apply the method of variable substitution leaves us with an x in the integrand, which makes the situation sufficiently complicated that it is worth exploring other alternatives:

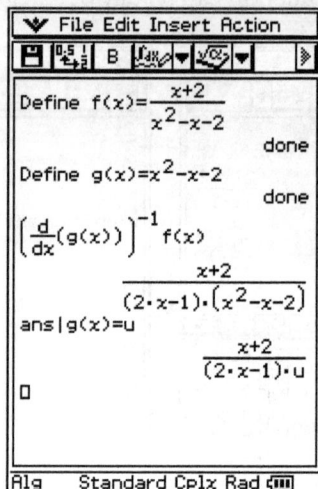

```
  ▼ File Edit Insert Action
 ▣ ▣ B ▣▱▾▱▾           ▷

Define f(x)= x+2
             ――――
             x²-x-2
                              done
Define g(x)=x²-x-2
                              done
 ⎛ d         ⎞⁻¹
 ⎜――(g(x))⎟    f(x)
 ⎝ dx       ⎠
               x+2
         ――――――――――――――
         (2·x-1)·(x²-x-2)
ans|g(x)=u
               x+2
            ――――――――
            (2·x-1)·u
□

Alg    Standard Cplx Rad ▣
```

A productive approach is to attempt to factor the denominator, then split the fraction into a sum of terms each of a simpler form:

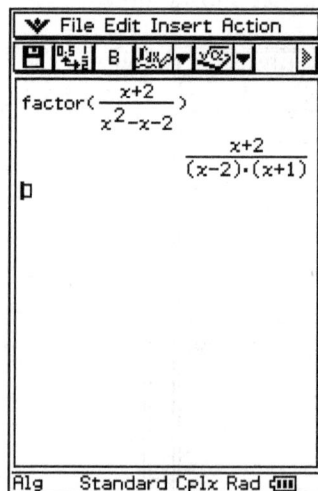

```
  ▼ File Edit Insert Action
 ▣ ▣ B ▣▱▾▱▾           ▷

factor( x+2 )
        ――――
        x²-x-2
                      x+2
                 ――――――――――――
                 (x-2)·(x+1)
□




















Alg    Standard Cplx Rad ▣
```

We see that the denominator is the product of 2 terms. We would like to find A and B so that:

$$\frac{A}{x-2} + \frac{B}{x+1} = \frac{x+2}{(x-2)(x+1)}$$

If we apply the combine() function to the left hand side of the expression, we will get an expression for the numerator in terms of A and B. Equating the coefficient of x and the constant term with the corresponding coefficients on the right hand side will give a pair of simultaneous equations for A and B.

Substituting these values in the left side of the above equation gives an integral we can all handle. (Of course ClassPad can handle the original equation, but it does not show its work as well as we do.)

## EXERCISES

Evaluate the following. Check your answers by differentiating in ClassPad.

1. $\displaystyle\int \frac{3}{2x-5}\,dx$

2. $\displaystyle\int \frac{3x-5}{x+1}\,dx$

3. $\displaystyle\int \frac{x+1}{x^2+2x+3}\,dx$

4. $\displaystyle\int \frac{3x-1}{x^2-1}\,dx$

5. $\displaystyle\int \frac{x^2+x+1}{x^2+2x-3}\,dx$

# Index

www.ingramcontent.com/pod-product-compliance
Lightning Source LLC
Chambersburg PA
CBHW081501200326
41518CB00015B/2345